Christian Gobert

Large Eddy Simulation of Particle-Laden Flow

Christian Gobert

Large Eddy Simulation of Particle-Laden Flow

Südwestdeutscher Verlag für Hochschulschriften

Impressum/Imprint (nur für Deutschland/ only for Germany)
Bibliografische Information der Deutschen Nationalbibliothek: Die Deutsche Nationalbibliothek verzeichnet diese Publikation in der Deutschen Nationalbibliografie; detaillierte bibliografische Daten sind im Internet über http://dnb.d-nb.de abrufbar.

Alle in diesem Buch genannten Marken und Produktnamen unterliegen warenzeichen-, marken- oder patentrechtlichem Schutz bzw. sind Warenzeichen oder eingetragene Warenzeichen der jeweiligen Inhaber. Die Wiedergabe von Marken, Produktnamen, Gebrauchsnamen, Handelsnamen, Warenbezeichnungen u.s.w. in diesem Werk berechtigt auch ohne besondere Kennzeichnung nicht zu der Annahme, dass solche Namen im Sinne der Warenzeichen- und Markenschutzgesetzgebung als frei zu betrachten wären und daher von jedermann benutzt werden dürften.

Verlag: Südwestdeutscher Verlag für Hochschulschriften Aktiengesellschaft & Co. KG
Dudweiler Landstr. 99, 66123 Saarbrücken, Deutschland
Telefon +49 681 37 20 271-1, Telefax +49 681 37 20 271-0
Email: info@svh-verlag.de
Zugl.: München, TU, Diss., 2010

Herstellung in Deutschland:
Schaltungsdienst Lange o.H.G., Berlin
Books on Demand GmbH, Norderstedt
Reha GmbH, Saarbrücken
Amazon Distribution GmbH, Leipzig
ISBN: 978-3-8381-1720-1

Imprint (only for USA, GB)
Bibliographic information published by the Deutsche Nationalbibliothek: The Deutsche Nationalbibliothek lists this publication in the Deutsche Nationalbibliografie; detailed bibliographic data are available in the Internet at http://dnb.d-nb.de.

Any brand names and product names mentioned in this book are subject to trademark, brand or patent protection and are trademarks or registered trademarks of their respective holders. The use of brand names, product names, common names, trade names, product descriptions etc. even without a particular marking in this works is in no way to be construed to mean that such names may be regarded as unrestricted in respect of trademark and brand protection legislation and could thus be used by anyone.

Publisher: Südwestdeutscher Verlag für Hochschulschriften Aktiengesellschaft & Co. KG
Dudweiler Landstr. 99, 66123 Saarbrücken, Germany
Phone +49 681 37 20 271-1, Fax +49 681 37 20 271-0
Email: info@svh-verlag.de

Printed in the U.S.A.
Printed in the U.K. by (see last page)
ISBN: 978-3-8381-1720-1

Copyright © 2010 by the author and Südwestdeutscher Verlag für Hochschulschriften Aktiengesellschaft & Co. KG and licensors
All rights reserved. Saarbrücken 2010

Abstract

This thesis is a contribution to research in the field of Large Eddy Simulation (LES) of particle-laden flow with a focus on the effect of unresolved small scale turbulence on suspended particles. The first substantial contribution of this thesis is a detailed quantification of small scale effects. The thesis contains new results from Direct and Large Eddy Simulation of particle-laden forced homogeneous isotropic turbulence. Reynolds numbers based on the Taylor length scale are $Re_\lambda = 34$, 52 and 99. Stokes numbers based on the Kolmogorov time scale range from $St = 0.1$ to $St = 100$. The main conclusions of these numerical experiments are the following: if subgrid scales are neglected for particle transport in LES, then particle kinetic energy is underpredicted, particle dispersion is overpredicted but preferential concentration is predicted satisfactorily.

The second contribution of this thesis is an analytical and numerical analysis of three commonly used models that describe the effect of subgrid scales on particles: the Approximate Deconvolution Method (ADM) and two stochastic models. Kuerten (Phys. Fluids 18, 2006) proposed ADM for particle-laden flow. The stochastic models were proposed by Shotorban & Mashayek (J. Turbul. 7, 2006) and Simonin et al. (Appl. Sci. Res. 51, 1993). Analytical and numerical results show that ADM improves the particle dynamics, but for coarse LES, the improvement is very small. On the other hand, at high Stokes numbers the predictions from the stochastic models are less accurate than those obtained by neglecting subgrid scales for particle transport. Furthermore, the stochastic models were found to destroy preferential concentration, whereas ADM preserves preferential concentration. In conclusion, the stochastic models were found to perform less reliably than ADM.

The third contribution of this thesis is a new model that can be regarded as an extension of ADM. The new model consists of a specific interpolation method, which is designed such that statistically the numerical interpolation error can be identified with the effect of the unresolved scales. The new model was assessed analytically and numerically. For the numerical assessment, simulations of forced homogeneous isotropic turbulence at $Re_\lambda = 52$, 99 and 265 were conducted. Analytical and numerical assessments show very promising results. In particular, the overall accuracy of the model is higher than the accuracy of ADM. As the coarseness of LES increases, the gain in terms of accuracy of the new model in comparison to ADM also increases. This means that for high Reynolds number configurations, where only coarse LES is possible, the new model can be expected to produce significantly better results than ADM.

Kurzfassung

Die vorliegende Arbeit stellt einen Forschungsbeitrag zum Thema Large Eddy Simulation partikelbeladener Strömungen dar. Der Schwerpunkt der Arbeit liegt auf dem Effekt der nicht aufgelösten kleinskaligen Turbulenz auf suspendierte Partikel. Der erste wesentliche Beitrag dieser Arbeit zur aktuellen Forschung ist eine detaillierte Quantifizierung kleinskaliger Effekte. Die Arbeit enthält neue Ergebnisse aus DNS und LES partikelbeladener homogener isotroper Turbulenz mit konstanter Energiezufuhr. Die Reynoldszahlen betragen $Re_\lambda = 34$, 52 und 99, basierend auf dem Taylorschen Längenmaß. Die Stokeszahlen reichen von $St = 0.1$ bis $St = 100$, basierend auf dem Komogorov'schen Zeitmaß. Die wichtigsten Ergebnisse dieser numerischen Experimente lauten wie folgt: Falls in LES die nicht aufgelösten Skalen für den Partikeltransport nicht berücksichtigt werden, dann wird die kinetische Energie der Partikel unterschätzt, die turbulente Dispersion überschätzt aber preferential concentration zufriedenstellend vorhergesagt.

Der zweite Beitrag dieser Arbeit ist eine analytische und numerische Analyse von drei üblicherweise eingesetzten Modellen für den Effekt der nicht aufgelösten Skalen auf die Partikel. Die drei Modelle sind die Approximate Deconvolution Method (ADM) und zwei stochastische Modelle. Kuerten (Phys. Fluids 18, 2006) schlug den Einsatz von ADM für partikelbeladene Strömungen vor. Die stochastischen Modelle wurden durch Shotorban & Mashayek (J. Turbul. 7, 2006) und Simonin et al. (Appl. Sci. Res. 51, 1993) vorgeschlagen. Analytische und numerische Ergebnisse zeigen, dass ADM zu einer verbesserten Vorhersage der Partikeldynamik führt wobei für grob aufgelöste LES die Verbesserung sehr gering ist. Andererseits liefern die stochastischen Modelle bei hohen Stokeszahlen Vorhersagen, die weniger genau sind als die Vorhersagen, die man erhält, wenn man die nicht aufgelösten Skalen für den Partikeltransport negiert. Außerdem zeigte sich, dass die stochastischen Modelle preferential concentration zerstören wohingegen ADM preferential concentration erhält. Zusammenfassend zeigte sich, dass die stochastischen Modelle weniger zuverlässig sind als ADM.

Der dritte Beitrag dieser Arbeit ist ein neues Modell, das als Erweiterung von ADM betrachtet werden kann. Das neue Modell besteht aus einer spezifischen Interpolationsmethode, die so konstruiert ist, dass statistisch der numerische Interpolationsfehler mit dem Effekt der nicht aufgelösten Skalen identifiziert werden kann. Das neue Modell wurde mit analytischen und numerischen Mitteln bewertet. Für die numerische Bewertung wurden Simulationen homogener isotroper Turbulenz mit Energiezufuhr bei $Re_\lambda = 52$, 99 und 265 durchgeführt. Die analytische und die numerische Bewertung zeigen sehr vielversprechende Ergebnisse. Insgesamt ist die Genauigkeit des Modells höher als die Genauigkeit von ADM. Je gröber die LES, desto höher ist der Genauigkeitsgewinn im Vergleich zu ADM. Das bedeutet, dass für hohe Reynoldszahlen, wo nur grobe LES möglich ist, erwartet werden kann, dass das neue Modell signifikant bessere Ergebnisse als ADM liefert.

Vorwort (Preface)

Diese Dissertation entstand — nein, nicht nur im Rahmen meiner Tätigkeit als wissenschaftlicher Angestellter an der TU München, sondern, wie üblich, auch in meiner Freizeit. Deshalb möchte ich den größten Dank an meine Frau Anika aussprechen, die immer Verständnis gezeigt hat, wenn ich mich zum 'Doktorarbeiten' zurückzog. Sie hat meine Arbeit und Veröffentlichungen korrekturgelesen, sich in die Thematik hineinversetzt und mir durch ihr Talent für kritische Fragen geholfen, Klarheit in meine Arbeit zu bringen.

Sehr viel Unterstützung erhielt ich von der gesamten Arbeitsgruppe des Fachgebiets Hydromechanik, insbesondere von meinem Doktorvater Herrn Prof. Manhart. Ich hatte immer das Gefühl, dass er mich nach allen Kräften finanziell, fachlich und persönlichkeitsbildend unterstützte. Er verbrachte mit mir viel Zeit in Diskussionen über meine Ideen und Ergebnisse und setzt Impulse in die richtige Richtung, so dass ich das Ziel meiner Promotion erreichen konnte. Außerdem fand ich in ihm immer wieder einen Ansprechpartner, der mir hilfreiche Hinweise in Bezug auf die Planung meiner Karriere gab. Er unterstützte mich stets ganzheitlich als Person. Ich danke ihm sehr für diese unglaublich breite und tiefgründige Unterstützung.

Herrn Prof. Simeon habe ich die Begeisterung für die numerische Mathematik zu verdanken. Ich danke ihm desweiteren für die Übernahme des Koreferats und für die finanzielle Unterstützung während eines Jahres im Laufe meiner Promotion. Ich hatte immer das Gefühl, dass er mich in meiner Promotion fördert, obwohl er nicht mein Doktorvater ist. In diesem Rahmen danke ich der gesamten Arbeitsgruppe M2 unter Leitung von Herrn Prof. Rentrop. 'Die Mathematiker' nahmen mich herzlich in ihrer Mitte auf.

Ich danke Herrn Prof. Kuerten, dass er sich als dritter Gutachter zur Verfügung stellte. Ich danke ihm insbesondere für die Organisation eines sehr inspirierenden und motivierenden Workshops in Eindhoven.

Allen drei Gutachtern danke ich für ihre Offenheit bezüglich interdisziplinärer Arbeit. In diesem Zuge bedanke ich mich bei Herrn Prof. Rank für die Gründung der International Graduate School of Science and Engineering (IGSSE). Die interdisziplinäre Ausrichtung der IGSSE motivierte mich stark für diese interdisziplinäre Arbeit. Ich freue mich, dass Herr Prof. Rank sich dazu bereit erklärt hat, den Vorsitz für mein Dissertationsverfahren zu übernehmen.

Ein ganz besonderer Dank geht an meine Eltern, die mir alle Voraussetzungen für eine Promotion mitgegeben und meine Arbeit korrekturgelesen haben.

Ich danke Vincent Daneker für die Revision meiner Arbeit und seine wohldurchdachten Verbesserungsvorschläge.

Zu guter Letzt danke ich dem DAAD für die Finanzierung zweier Forschungsaufenthalte in Korea und Chile. Die fachlichen und interkulturellen Erfahrungen aus diesen Aufenthalten sind für mich unersetzlich und ich bin mir sicher, dass ich von diesen Erfahrungen nicht nur während meiner Promotion sondern auch während meiner zukünftigen Karriere profitieren werde.

Contents

Table of contents v
List of tables viii
List of figures ix
Nomenclature xiii

1 Introduction   1
  1.1 Motivation and background of the work 1
  1.2 Modelling tasks for Large Eddy Simulation of particle-laden flow 3
  1.3 Objective of this work 3
  1.4 Outline of the work 3

2 Fundamentals of particle-laden flow   5
  2.1 Single phase turbulent flow 5
    2.1.1 Navier–Stokes equations and turbulence 6
    2.1.2 Richardson's and Kolmogorov's theories on turbulence 7
    2.1.3 Statistical and spectral description of isotropic turbulence 9
  2.2 Particle-laden turbulent flow 14
    2.2.1 Classification and specification 15
    2.2.2 Maxey–Riley equation and its history 19
    2.2.3 Turbulent dispersion and preferential concentration 24

3 Implemented simulation tools   28
  3.1 Simulation techniques for turbulent flow 28
    3.1.1 Direct Numerical Simulation (DNS) 28
    3.1.2 Large Eddy Simulation (LES) 29
  3.2 Numerical methods for the carrier fluid 36
    3.2.1 Basic numerical scheme 36
    3.2.2 Forcing of isotropic turbulence 37
  3.3 Numerical methods for particle-laden flow 38
    3.3.1 Interpolation of the fluid velocity on the particle's position 38
    3.3.2 Computation of particle velocity 40
  3.4 Summary of implemented numerical methods 42

4 A methodology for assessment of particle-LES models   43
  4.1 Definition of an averaging operator for particles 43
  4.2 Definition of assessment criteria 46
  4.3 A priori and a posteriori analysis 47
  4.4 Conclusions of chapter 4 49

5 A numerical study on requirements for a particle-LES model   51
    5.1   Relation to previous works 51
    5.2   Numerical Simulation of the carrier flow 52
    5.3   Parameters for the discrete particle simulations 57
    5.4   Validation 58
    5.5   Effect of the SGS turbulence on the particles 58
         5.5.1   Kinetic energy. 58
         5.5.2   Integral time scale 67
         5.5.3   Particle dispersion 73
         5.5.4   Preferential concentration 75
    5.6   Conclusions of chapter 5 76

6 Presentation and assessment of existing particle-LES models   78
    6.1   A word on notation in this chapter 78
    6.2   Particle-LES models 78
         6.2.1   Overview on particle-LES models 79
         6.2.2   Analysed particle-LES models 79
    6.3   Analytical assessment 82
         6.3.1   Framework for the analytical computations 82
         6.3.2   First moments. 83
         6.3.3   Second moments 90
    6.4   Numerical assessment 96
         6.4.1   Numerical results from literature 96
         6.4.2   Numerical assessment of ADM 97
         6.4.3   Numerical assessment of the Langevin-based models 105
         6.4.4   Preferential concentration 109
    6.5   Conclusions of chapter 6 109

7 A Novel Particle-LES Model based on Spectrally Optimised Interpolation (SOI) 112
    7.1   Preliminary considerations 112
         7.1.1   The spectrum seen by particles 113
         7.1.2   ADM revisited - ADM and interpolation 119
    7.2   Construction of the SOI model 122
         7.2.1   Properties of the model (admissibility conditions) 123
         7.2.2   Formulation of the model as optimisation problem 124
         7.2.3   Reduction of computational overhead: Optimisation against 1D-spectra 125
         7.2.4   Imposing admissibility conditions 128
         7.2.5   The SOI model formulated as cooking recipe 131
    7.3   Analytical assessment of the SOI model 134
    7.4   Numerical assessment of the SOI model 138
         7.4.1   Interpolation stencils 138
         7.4.2   One-dimensional Spectra 139
         7.4.3   Particle dynamics 142
    7.5   Relation of SOI to other models and outline of extensions for arbitrary flow . 147
    7.6   Conclusions of chapter 7 149

8 Conclusions 151
8.1 Summary of results 151
8.2 Possible extensions of this thesis 153

Bibliography 155

Appendix: Computational requirements for SOI 168

List of Tables

2.1	Classification of particle-laden flow in view of numerical simulation with respect to an appropriate mathematical description.	16
2.2	Classification of particle-laden flow with respect to equation coupling following Elghobashi (1991).	17
5.1	Simulation parameters and Eulerian statistics from DNS of forced isotropic turbulence.	53
5.2	Parameters for LES of forced isotropic turbulence and time scale of energy containing eddies $[k_f]/[\epsilon]$ computed from resolved scales.	57
5.3	Kolmogorov constant C_0 following Fox & Yeung (2003) (equation (5.21)). . .	68
7.1	Integral time scales computed from the model presented in section 7.3 for SOI, ADM and LES without particle-LES model.	138
7.2	Qualitative summary of the results from section 7.4.3.	147
A.1	Computational costs for computation of fluid velocity at particle position in terms of floating point operations (flops).	168

List of Figures

1.1 Aerosol pollution over Northern India and Bangladesh and Mississippi River Sediment Plume. 2

2.1 Longitudinal and transverse two point correlation functions. Computed from the model spectrum proposed by Pope (2000) for $Re_\lambda = 34$. 11
2.2 Model spectrum proposed by Pope (2000) for $Re_\lambda = 52$, $Re_\lambda = 99$ and $Re_\lambda = 280$. 13
2.3 Flow behind a sphere. Photographs taken by Taneda (1956) (reprinted with permission). 22
2.4 Drag coefficient of a sphere as a function of particle Reynolds number according to Clift et al. (1978) and Schiller & Naumann (1933). 22
2.5 DNS of Lycopodium in turbulent channel flow (air), instantaneous distribution on channel centreplane, $Re_\tau = 180$, $St = 0.6$. 26
2.6 DNS of copper particles in turbulent channel flow (air), instantaneous distribution on channel centreplane, $Re_\tau = 180$, $St = 56$. 26

3.1 Filter transfer functions of sharp spectral, box and Gaussian filter. 34
3.2 Filter transfer function of the Lagrangian Smagorinsky model from section 3.1.2 together with the transfer function of the box filter and the Gaussian filter. 35

4.1 Schematic of steps in the a priori analysis. 49

5.1 Instantaneous energy spectrum functions from DNS. 55
5.2 Instantaneous energy spectrum functions from DNS and Pope's model. ... 55
5.3 Instantaneous energy spectrum functions from LES. 56
5.4 Instantaneous energy spectrum functions from DNS, filtered DNS and LES. 56
5.5 Probability density function of particle acceleration. 58
5.6 A priori analysis: Unfiltered and filtered kinetic energy of the fluid seen by the particles computed from DNS. 59
5.7 A posteriori analysis: Kinetic energy seen by the particles in DNS and LES. 59
5.8 A priori and a posteriori analysis: Kinetic energy of the fluid seen by the particles, scaled by resolved kinetic energy. 60
5.9 A priori analysis: Kinetic energy of the fluid seen by the particles computed from DNS, scaled by resolved kinetic energy. Comparison of different Reynolds numbers. 61
5.10 A posteriori analysis: Kinetic energy of the fluid seen by the particles, scaled by resolved kinetic energy and smallest resolved time scale. 61

5.11 A priori analysis: Kinetic energy of the fluid seen by the particles, scaled by resolved kinetic energy and large eddy decay time. 62
5.12 A posteriori analysis: Kinetic energy of the fluid seen by the particles, scaled by resolved kinetic energy and resolved large eddy decay time. 62
5.13 A posteriori analysis: Kinetic energy of the fluid seen by the particles, scaled by resolved kinetic energy and resolved large eddy decay time. 63
5.14 A priori analysis: Covariance of filtered velocity and fluctuations seen by the particles. 65
5.15 A priori analysis: Kinetic energy of the fluctuations seen by the particles. . . 65
5.16 A posteriori analysis: Kinetic energy of the particles together with analytical estimates. 66
5.17 A posteriori analysis: Kinetic energy of the particles with respect to the kinetic energy of the fluid seen by the particles together with analytical estimates. . 67
5.18 A priori analysis: Integral time scale of unfiltered and filtered fluid velocity seen by the particles. 68
5.19 Integral time scale of the fluid velocity seen by the particles in DNS, filtered DNS and LES. 69
5.20 Autocorrelation of fluid velocity seen by particles at $Re_\lambda = 99, St = 0.1$. . . . 70
5.21 A priori analysis: Integral time scale of the fluid seen by the particles computed from DNS, scaled by $t_{u@p}(St = 1)$ and $\langle t_{u@p}(St = 1)\rangle$, respectively. . . 71
5.22 A posteriori analysis: Integral time scale of the fluid seen by the particles, scaled by $t_{u@p}(St = 1)$ and $[t_{u@p}(St = 1)]$, respectively. 71
5.23 A priori analysis: Integral time scale of the fluid seen by the particles computed from DNS, scaled by $t_{u@p}(St_L = 1)$ and $\langle t_{u@p}(St_L = 1)\rangle$, respectively. . 72
5.24 A posteriori analysis: Integral time scale of the fluid seen by the particles, scaled by $t_{u@p}(St_L = 1)$ and $[t_{u@p}(St_L = 1)]$, respectively. 72
5.25 A priori analysis: Integral time scale of the small scale fluid velocity seen by the particles. 73
5.26 A posteriori analysis: Integral time scale of particle velocity. 74
5.27 A posteriori analysis: Turbulent particle dispersion. 74
5.28 A posteriori analysis of preferential concentration. 75

6.1 Sketch for modelling errors dependent on particle relaxation time and wavenumber. 89
6.2 Sketch for modelling errors for $\tau_p > \tau_c$. 89
6.3 Spectrum from DNS and model spectrum propsed by Pope (2000) for $Re_\lambda = 52$. 101
6.4 ADM stencil for $Re_\lambda = 52$ obtained by optimisation against DNS spectrum (ADM^{DNS}). 101
6.5 Transfer function of ADM stencil shown in figure 6.4, LES transfer function $\sqrt{E^{LES}/E^{DNS}}$ obtained a posteriori from DNS and LES of isotropic turbulence at $Re_\lambda = 52$ and product of both transfer functions. 101
6.6 ADM stencil for $Re_\lambda = 52$ obtained by Kuerten's approach and by optimisation against spectra. 102
6.7 Transfer function of ADM stencils shown in figure 6.6. 102
6.8 A priori and a posteriori analysis of ADM, kinetic energy seen by particles (second moment in fluid velocity seen by particles). 103

List of Figures

6.9 A priori and a posteriori analysis of ADM, particle kinetic energy (second moment in particle velocity). 104
6.10 A priori and a posteriori analysis of ADM, rate of dispersion (second moment in particle position). 104
6.11 A priori and a posteriori analysis of the Langevin-based models, kinetic energy seen by particles (second moment in fluid velocity seen by particles). 107
6.13 A priori and a posteriori analysis of the Langevin-based models, rate of dispersion (second moment in particle position). 108
6.12 A priori and a posteriori analysis of the Langevin-based models, particle kinetic energy (second moment in particle velocity). 108
6.14 A posteriori analysis of preferential concentration. 109

7.1 1D-spectra of isotropic turbulence at $Re_\lambda = 52$ 114
7.2 Interpolation kernels for piecewise linear and cubic interpolation. 117
7.3 Transfer functions for piecewise linear and cubic interpolation. 117
7.4 Longitudinal and transverse spectra seen by particles in isotropic turbulence at $Re_\lambda = 52$, computed by DNS. 118
7.5 Longitudinal and transverse spectra seen by particles in isotropic turbulence at $Re_\lambda = 52$, computed by LES. 119
7.6 Longitudinal and transverse spectra seen by particles in isotropic turbulence at $Re_\lambda = 52$, computed by LES with and without ADM 121
7.7 Longitudinal and transverse spectra seen by particles in isotropic turbulence at $Re_\lambda = 52$, computed by LES with ADM, fourth- and second-order interpolation. 121
7.8 Longitudinal and transverse spectra seen by particles in isotropic turbulence at $Re_\lambda = 52$, computed by LES with ADM. Comparison of results from ADM, optimised against DNS spectrum and ADM, optimised against model spectrum. 122
7.9 Samples of the functions a, b, C_1^{SOI} and C_2^{SOI}. 127
7.10 Example showing that the equality of Eulerian spectra does not necessarily entail equality of the Lagrangian spectra. 135
7.11 The distribution F from equation (7.59) for fixed κ and for fixed ω 135
7.12 Lagrangian spectra computed from the model spectrum of section 2.1.3 for $Re_\lambda = 265$ and equation (7.61). 137
7.13 Autocorrelation functions that correspond to the spectra plotted in figure 7.12. 137
7.14 Stencils for ADM and SOI and corresponding transfer functions. 139
7.15 Longitudinal and transverse spectra seen by particles in isotropic turbulence at $Re_\lambda = 52$, computed by LES with ADM and SOI. ADM and SOI stencils were obtained by optimisation against the DNS spectrum. 140
7.16 Longitudinal and transverse spectra seen by particles in isotropic turbulence at $Re_\lambda = 52$, computed by LES with ADM and SOI. ADM and SOI stencils were obtained by optimisation against a model spectrum. 141
7.17 Longitudinal and transverse spectra seen by particles in isotropic turbulence at $Re_\lambda = 99$, computed by LES with ADM and SOI. 141
7.18 Longitudinal and transverse spectra seen by particles in isotropic turbulence at $Re_\lambda = 265$, computed by LES with ADM and SOI. 141
7.19 Kinetic energy of the fluid seen by the particles in DNS, LES, ADM and SOI. 143
7.20 Kinetic energy of the particles in DNS, LES, ADM and SOI. 143

7.21 Kinetic energy of the particles in DNS, LES, ADM and SOI. Zoomed-in view of figure 7.20 144
7.22 Integral time scale of the fluid velocity seen by the particles in DNS, LES, ADM and SOI. 145
7.23 Integral time scale of particle velocity in DNS, LES, ADM and SOI. 145
7.24 Rate of dispersion in DNS, LES, ADM and SOI. 146
7.25 Preferential concentration at $Re_\lambda = 52$. Accumulation Σ and fractal dimension d_{pc}. Results from DNS, LES, ADM and SOI. 146

Nomenclature

---------- Roman letters ----------

C_0^{SOI} .... model parameter for SOI
C_1^{SOI}, C_2^{SOI} ...... model functions for SOI
c_D .... drag coefficient
C_S .... Smagorinsky constant
d_{pc} .... fractal dimension
D .... rate of dispersion
d .... particle diametre
e^{ADM} .... error in first moments for ADM
E^{mod} .... modelled energy spectrum function
e^{Sho} .... error in first moments for the model of Shotorban & Mashayek (2006)
e^{Sim} .... error in first moments for the model of Simonin et al. (1993)
$e_i(\kappa)$ .... sensitivity of the error in velocity component v_i
$E_l^{target}, E_t^{target}$ ... longitudinal or transverse target spectra for SOI, either computed from DNS or model spectra
E_l .... longitudinal spectrum
E_t .... transverse spectrum
E_{ij} .... one-dimensional spectra
E_{Lag} .... Lagrangian frequency spectra
E .... energy spectrum function
G .... filter kernel
$\mathbf{g}$ .... gravity
k, k_f .... kinetic energy
k_p .... kinetic energy of particles
k_{sgs} .... subgrid kinetic energy
$k_{u@p}$ .... kinetic energy seen by particles
$\mathbf{k}$ .... wavenumber
L_f .... longitudinal integral length scale
L_k .... length scale of energy containing eddies
N_p .... number of particles
$p(\mathbf{x}, t)$ .... pressure of the carrier fluid
Re .... Reynolds number
Re_λ .... Reynolds number based on λ and u_{rms}
Re_{L_k} .... Reynolds number based on L_k and k
Re_p .... particle Reynolds number based on d and $\|\mathbf{u}_{f@p} - \mathbf{u}_p\|$
St .... Stokes number based on Kolmogorov time scale

$\mathbf{S}$	rate-of-strain tensor
T_L	time scale for the model of Shotorban & Mashayek (2006)
t_p	integral time scale of particle velocity
$t_{u@p}$	integral time scale of fluid velocity seen by particles
u_{rms}	rms velocity
$\mathbf{u}_f(\mathbf{x},t)$	velocity of the carrier fluid
$\mathbf{u}_p(t)$	particle velocity
$\mathbf{u}_{f@p}(t)$	fluid velocity seen by the particle
$\mathbf{v}$	$\mathbf{v} = \mathbf{u}_f$ for the analysis of the model of Shotorban & Mashayek (2006), $\mathbf{v} = \mathcal{H}\mathbf{u}_f$ for the analysis of the model of Shotorban & Mashayek (2006)
w_l, w_t	longitudinal and transverse 1D kernel for SOI
w_{lin}	1D kernel for linear interpolation
$\mathbf{W}$	Wiener process
$\mathbf{w}, \mathbf{W}$	weighting function, interpolation kernel
$\mathbf{W}_{cl}, w_{cl}$	3D and 1D kernel for second-order conservative interpolation
$\mathbf{W}_{cub}, w_{cub}$	3D and 1D kernel for fourth-order (i.e. cubic) interpolation
$\mathbf{W}_{sl}, w_{sl}$	3D and 1D kernel for semi-linear interpolation
$\mathbf{W}_{SOI}$	3D kernel for SOI
$\mathbf{W}_{tl}$	3D kernel for trilinear interpolation
$\mathbf{x}_p(t)$	particle position

--- **Greek letters** ---

$\delta_{a,b}$	Dirac delta-function
δ_{ij}	Kronecker delta-function
Δ	filter width
δ	error bound
ϵ	rate of dissipation of kinetic energy
η_K	Kolmogorov length scale
Γ	parameter for the model of Simonin *et al.* (1993)
κ_c	cutoff wavenumber
κ	norm of wavenumber
λ	transverse Taylor length scale
ν_t	eddy viscosity
ν	kinematic viscosity of the carrier fluid
ρ, ρ_f	density of the carrier fluid
ρ_p	material density of particle
Σ	measure for accumulation
τ_K	Kolmogorov time scale
τ_p	particle relaxation time
τ	subgrid stress tensor

Nomenclature

Operators

$\cdot^*$	complex conjugate
$\overline{\cdot}$	ensemble averaging
$\cdot'$	turbulent fluctuations / SGS quantity
$\langle \cdot \rangle$	filtered quantity
$[\cdot]$	LES quantity
$\|\cdot\|$	2-norm
$\cdot_{@p}$	quantity evaluated at particle position
$\mathcal{FT}$	Fourier transformation
$\mathcal{G}$	spatial filter
$\mathcal{H}$	extractor of subgrid scales, $\mathcal{H} = \mathcal{I} - \mathcal{G}$

In general, the Einstein notation is employed, e.g. $\frac{\partial u_i}{\partial x_i} = \sum_i \frac{\partial u_i}{\partial x_i}$, $S_{ij}S_{ij} = \sum_i \sum_j S_{ij}S_{ij}$

Abbreviations

1D	one dimension
3D	three dimensions
ADM	Approximate Deconvolution Method
ADM^{DNS}	ADM based on optimisation against DNS spectrum
ADM^{Kuerten}	ADM following Kuerten (2006b)
ADM^{mod}	ADM based on optimisation against model spectrum
DNS	Direct Numerical Simulation
LES	Large Eddy Simulation
rms	root-mean square
SGS	subgrid scale(s)
Sho	refers to the model proposed by Shotorban & Mashayek (2006)
Sim	refers to the model proposed by Simonin et al. (1993)
SOI	Spectrally Optimised Interpolation

Terminology

fluid velocity seen by the particle	The fluid velocity interpolated at the particle's position, i.e., $\mathbf{u}_{f@p}(t)$
fluid-LES model	model for the effect of unresolved scales of the fluid flow on the resolved scales
particle-LES model	model for the effect of unresolved scales of the fluid flow on the particles

1 Introduction

'Turbulence modelling is dirty work' I was once told by a professor for mathematics. 'People tend to call crude assumptions a model and then they 'prove' that the model works by applying it on two or three test cases which they chose by themselves'. Well, today I would translate this statement by 'turbulence modelling is not easy and in order to develop a model for realistic problems, it is sometimes unavoidable to assume some rather weakly justifiable relations. If these assumptions lead to good results for a specific test case, then the model is suited for configurations which are similar to that test case.' Thus, during my time as PhD student I learned that turbulence modelling is not dirty work but a great challenge.

1.1 Motivation and background of the work

The present work focuses on particle-laden flow. Typical background applications are, for example, sedimentation and deposition of aerosols. Figure 1.1 shows the Mississippi river sediment plume and aerosol pollution over Northern India and Bangladesh. The aerosol pollution is a result of human activity (according to NASA the aerosol is rich in sulfates, nitrates, organic and black carbon, and fly ash), leading to environmental hazards. The Mississippi sediment plume also has a significant impact on the environment. NASA states that it extends the coastline by 91m per year. Such problems form the motivation for this work. The flows under consideration are particle-laden turbulent flows. The work is a contribution to the challenging task of numerical prediction of such flows.

Particle-laden flow can also be found in other disciplines such as medical science (e.g. deposition in the respiratory tract) or chemical engineering (e.g. chemical precipitation). The wide range of applications makes it interesting to develop a general method for numerical prediction of particle-laden flow in contrast to specialised methods for sedimentation, aerosol deposition, etc.

In the last decades several such methods were developed. Recently, Guha (2008) reviewed state of the art computational methods for this field. If direct numerical simulation (DNS) is possible, then state of the art methods for dilute particle-laden flow produce reliable results (cf. e.g. Geurts *et al.*, 2007; Balachandar & Eaton, 2010). DNS means solving the Navier–Stokes equations 'as is'. This approach is very exact but computational requirements are immense. An alternative approach is Large Eddy Simulation (LES). Here, one solves for the large scales only. This method is less exact than DNS but the computational requirements are very much lower. The present work focuses on LES of particle-laden flow.

Often people asked me whether this means that if new, faster computers will be developed then my work will become meaningless because then particle-laden flow can be computed by

Figure 1.1: Left: Aerosol pollution over Northern India and Bangladesh, top: Mississippi River Sediment Plume. Source: NASA, Visible Earth, cf. http://visibleearth.nasa.gov

DNS. In order to answer that question, let us consider again the example of sediment transport in the Mississippi river. According to U.S. Geological Survey (field measurement data available online at http://waterdata.usgs.gov), the Mississippi shows at Baton Rouge an average flow velocity of about 0.9m/s, a maximum depth of 9m and a width of 1km. With this data, one can estimate that with today's computers the computation of the flow at Baton Rouge by DNS requires approximately 7.5×10^{16} CPU hours[1]. For comparison: if one would use all 9728 cores of the present supercomputer HLRB2 in Garching, Germany, then 7.5×10^{16} CPU hours would still mean 882Mio. years of non-stop computing. But this is only the flow at Baton Rouge. On top of that, the river basin must be computed. These requirements greatly outstrip the computational capacities of near future computers.

---

[1] Details of the estimate following Reynolds (1990):
Reynolds number based on flow depth and average velocity: $8.1 \cdot 10^6$
Resolution requirements from Reynolds (1990)'s estimates for turbulent (half) channel flow, grid stretched towards open surface:
Height of domain: 9m, smallest scales: $9m/Re^{0.9} = 5.5 \cdot 10^{-6} m$
Length of domain: 18m
Resolve complete transverse length (inhomogeneous): 1km
Number of cells vertical: $64(Re/3300)^{0.9} = 72000$
Number of cells longitudinal: $129(Re/3300)^{0.9} = 145000$
Number of cells transverse: $1km/5.5 \cdot 10^{-6} m = 1.8 \cdot 10^8$
Time step size at $CFL = 1 : 5.5 \cdot 10^{-6} s$ (average velocity: $0.9 m/s$)
Simulation time: 10 flow through times, $36 \cdot 10^6$ time steps
Performance: 4 CPU seconds per 10^6 cells per time step
Overall: $2.7 \cdot 10^{20}$ CPU seconds

1.2 Modelling tasks for Large Eddy Simulation of particle-laden flow

The immense computational requirements show the need for methods for LES of particle-laden flow. This is addressed in the present work. As mentioned above, the idea of LES is to solve for the large scales only.

In order to understand the modelling tasks in LES, one can assume that the small unresolved scales would be known as well. One can think of the large scales as large vortices and the small scales as small vortices. It is clear that if one generates additional small vortices, for example by small stirring motions, then these small vortices will affect the large vortices. This shows an intrinsic problem of LES. The unresolved small scales have an effect on the resolved large scales. This effect manifests itself in the non linear term of the Navier–Stokes equations. Thus, already for single phase flow, LES needs a model for this effect. Such models are hereafter referred to as fluid-LES models.

There are many studies proposing and assessing fluid-LES models, pointing out the respective advantages and disadvantages, limits and capabilities (see e.g. Sagaut, 2006). Nevertheless, this field is very challenging and there still remain many open problems, such as the treatment of complex geometries or flow detachment. Present research (see Brun, Juvé, Manhart & Munz, 2009) shows promising progress in tackling the remaining problems, making LES reliable for arbitrary single phase flows. Given this background, the present work focuses on LES of *particle-laden* flow.

For LES of particle-laden flow, one needs to respect that the unresolved scales of the carrier flow have an effect on the particles. Again, modelling is necessary. In the present work these models are referred to as particle-LES models.

1.3 Objective of this work

There is little work available so far concerning quantification and modelling of small scale turbulence effects on particles. The present work attempts to fill this gap by numerical experiment and analytical considerations.

In the present work, physical mechanisms that a particle-LES model must emulate are identified. Furthermore, existing models are assessed with respect to their capability to emulate these mechanisms. Based on these observations, a new particle-LES model is proposed. The model is based on a completely new idea, making use of numerical errors caused by interpolation in order to model small scale effects. The model is assessed using the same numerical and analytical methods that were applied previously on the other models.

1.4 Outline of the work

The present work is organised as follows. Chapter 2 covers some fundamental issues concerning particle-laden flow. Chapter 3 focuses on numerical tools. Simulation techniques and numerical methods which are implemented in this work are presented. These two chapters represent a selection of formerly published results.

Chapters 4 to 7 cover new results concerning LES of particle-laden flow. Chapter 4 contains the methodology which is the basis of the remaining work. Chapter 5 investigates the effect of small scale turbulence on particles by numerical experiments. This effect must be emulated by a particle-LES model. Therefore chapter 5 defines requirements for such a model. Chapter 6 analyses existing particle-LES models with respect to that. The chapter focuses on three of the most promising particle-LES models. The analysis shows that all three models contain significant structural defects. Finally, a new particle-LES model is presented in chapter 7 which does not contain such defects.

2 Fundamentals of particle-laden flow

The present chapter contains a brief literature review on single phase and particle-laden turbulent flow. The history of research on turbulent flow can be traced back to Leonardo da Vinci who wrote:

> *Observe the motion of the surface of the water, which resembles that of hair, which has two motions, of which one is caused by the weight of the hair, the other by the direction of the curls; thus the water has eddying motions, one part of which is due to the principal current, the other to random and reverse motion.*
>
> (Leonardo da Vinci, , circa 1500)

Maybe it was this inspiring quotation which motivated researchers since then to publish a seemingly endless series of studies on turbulent flow. In view of this, the present chapter can only be seen as an attempt to extract those findings which are most relevant for the present thesis.

Section 2.1 focuses on single phase turbulent flow. The Navier–Stokes equations are presented and the theories of Richardson and Kolmogorov are summarised. In addition, section 2.1 contains a description of isotropic turbulence. Isotropic turbulence will be used throughout this thesis as reference testcase.

Section 2.2 contains a literature review on particle-laden flow. This section also contains a precise specification of the type of particle-laden flow which is considered in the present thesis. Governing equations are presented and physical effects of particles in isotropic turbulence are explained.

Numerical approaches are not discussed in the present chapter but rather in chapter 3. In any case, high performance computing surely was not relevant for Leonardo's work.

2.1 Single phase turbulent flow

This section covers some fundamental aspects of single phase turbulent flow. The focus is on those aspects which are relevant to the subsequent chapters. For more comprehensive literature on this topic the reader is referred to Batchelor (1982); Tennekes & Lumley (1972); Pope (2000); Landau & Lifshitz (1987); Rotta (1972); Frisch (1995) and Hinze (1975).

In the following, first the Navier–Stokes equations are presented and a definition of turbulence is given. Then, two basic theories on turbulent flow are presented, namely the theories of Richardson (1922) and Kolmogorov (1941). These theories concern the statistical description of turbulent flow. Finally, isotropic turbulence is defined and discussed.

2.1.1 Navier–Stokes equations and turbulence

In this work, incompressible Newtonian fluids, governed by the Navier–Stokes equations

$$\frac{\partial u_{f,i}}{\partial x_i} = 0 \qquad (2.1a)$$

$$\frac{\partial u_{f,i}}{\partial t} + u_{f,j}\frac{\partial u_{f,i}}{\partial x_j} = -\frac{1}{\rho}\frac{\partial p}{\partial x_i} + \nu\frac{\partial^2 u_{f,i}}{\partial x_j^2} \qquad (2.1b)$$

are considered. $\mathbf{u}_f(\mathbf{x}, t)$ is the velocity of the fluid, $p(\mathbf{x}, t)$ the pressure and ν denotes kinematic viscosity. For particle-laden flow, the Navier–Stokes equations describe the carrier flow.

Fluid flows can be classified in turbulent and laminar flow. According to Rotta (1972), turbulent flows

- are irregular,
- show vortices,
- are three-dimensional,
- are unsteady.

Only if a flow shows all these characteristics, then it is called turbulent. Otherwise, it is laminar.

Reynolds number

One of the first experiments on turbulence was carried out by Reynolds (1883, 1895). He studied the flow through a long pipe and observed that turbulence occurs if the Reynolds number Re is high enough. In general, the Reynolds number is defined by a length scale L, a velocity scale U and the viscosity ν

$$Re = \frac{LU}{\nu}. \qquad (2.2)$$

L and U are characteristic scales for the flow. For pipe flow, one typically sets L to the pipe diametre and U to the area-averaged axial velocity. A pipe flow is laminar if Re is less than approximately 2300 and turbulent if the Reynolds number is higher than 4000. In the range in between, transition occurs. In that range, the flow can be laminar but small perturbations of the flow lead to turbulence.

In general, the critical Reynolds number, i.e., the Reynolds number where transition occurs, depends on the configuration. For an arbitrary configuration the critical Reynolds number can only be determined by numerical or physical experiment.

Reynolds decomposition

Already Leonardo da Vinci observed that turbulent flows can be decomposed in two parts. He noted that 'one part of which is due to the principal current, the other to random and

2 Fundamentals of particle-laden flow

reverse motion', cf. page 5. Thus, one might say that Leonardo proposed to decompose a turbulent flow into mean flow and fluctuations.

However, in general one associates this decomposition with Reynolds (1895) because he was the first to state this decomposition mathematically. Following Reynolds, one writes

$$\mathbf{u}(\mathbf{x},t) = \bar{\mathbf{u}}(\mathbf{x},t) + \mathbf{u}'(\mathbf{x},t) \tag{2.3a}$$
$$p(\mathbf{x},t) = \bar{p}(\mathbf{x},t) + p'(\mathbf{x},t) \tag{2.3b}$$

$\bar{\mathbf{u}}$ and $\bar{p}$ denotes averaged velocity and pressure, $\mathbf{u}'$ and p' are called fluctuations.

For general flow, averaging means ensemble averaging, i.e., averaging over several realisations of the same experiment. It should be noted that in this case also the averaged velocity depends on space and time.

In the present work, the testcase for numerical simulation is isotropic turbulence, explained in section 2.1.3. As explained in that section, isotropic turbulence means that ensemble averaged quantities are independent of space. In addition, in the present work a scheme for constant application of energy on the flow is implemented. Such a scheme is called forcing scheme, cf. section 3.2.2. It guarantees that ensemble averaged quantities are independent of time. Concluding, in forced isotropic turbulence ensemble averaging corresponds to spatial and temporal averaging.

2.1.2 Richardson's and Kolmogorov's theories on turbulence

In the previous section the decomposition of a turbulent flow into mean velocity and fluctuations was presented. The logical next step consists in a statistical description of the fluctuations.

In the present section, two theories for the statistical description of a turbulent flow are presented. One is the theory of Richardson and the other the theory of Kolmogorov. Both theories aim at a description which holds for all turbulent flows, i.e., there is no specific geometry assumed.

Richardson's energy cascade

Richardson (1922) presented one of the first theories on turbulence. His idea was to decompose a flow in motions of various scales. He referred to such motions as eddies. Leonardo's description of a turbulent flow (cf. page 5) gives an idea of what an eddy is but there exists no precise mathematic definition. Pope (2000) writes that 'an eddy eludes precise definition'. However, the idea is to regard a flow as a superposition of some coherent rotating chunks, maybe vortices, such that with each chunk some length scale can be associated and that the length scale defines a typical velocity and life time of this chunk.

In practice, one often defines eddies from the spatial Fourier transform of the fluid velocity, $(\mathcal{FT}(\mathbf{u}_f))(\mathbf{k})$. The argument of the Fourier transform $\mathbf{k}$ is refered to as wavenumber. Small eddies contribute to the Fourier transform at high wavenumbers (i.e. large $\|\mathbf{k}\|$) and large eddies contribute at low wavenumbers (i.e. small $\|\mathbf{k}\|$). The ensemble of all large eddies are referred to as 'large scales' and the ensemble of all small eddies are referred to as 'small scales'.

$\mathcal{FT}(\mathbf{u}_f)$ is a function with three components and depends on the three-dimensional wavenumber $\mathbf{k}$. For easier presentation, it is useful to define the scalar energy spectrum function $E(\|\mathbf{k}\|)$ which depends on the scalar value $\|\mathbf{k}\|$,

$$E(\|\mathbf{k}\|) = \int_{\|\mathbf{k}'\|=\|\mathbf{k}\|} \|(\mathcal{FT}(\mathbf{u}_f))(\mathbf{k}')\|^2 \, d\mathbf{k}'. \tag{2.4}$$

One rigorous definition of turbulence is that if E is continuous, then the flow is called turbulent. It should be mentioned that the equivalence of this definition and the definition from section 2.1.1 was to date only shown heuristically (cf. Ruelle, 2003).

Richardson (1922) postulated that large eddies create small eddies. In other words, the energy $E(\kappa)$, contained in the scales of size $2\pi/\kappa$, leads to creation of eddies of size $\kappa' > \kappa$. This means that the energy is constantly transferred towards higher wavenumbers. Therefore Richardson's theory is also referred to as energy cascade.

Richardson furthermore postulated that there is a lower limit for the eddy size. Below that limit, no eddy can exist because it would be converted instantly into heat due to internal friction. This conversion of energy is referred to as dissipation (see e.g. Mathieu & Scott, 2000).

Concerning dissipation, one should mention the work of Taylor (1922). At about the same time as Richardson, Taylor published a work on statistical theory for turbulent flow. He considered averaged quantities such as average kinetic energy

$$k = \frac{1}{2}\overline{u_{f,i}^2}. \tag{2.5}$$

As above, $\overline{}$ denotes ensemble averaging. So far, k depends on time. Now, one can derive a differential equation for the temporal development of k and one will find a sink term ϵ in this equation (see e.g. Pope, 2000),

$$\epsilon = 2\nu\overline{S_{ij}S_{ij}}, \qquad S_{ij} = \frac{1}{2}\left(\frac{\partial u_{f,i}}{\partial x_j} + \frac{\partial u_{f,j}}{\partial x_i}\right). \tag{2.6}$$

ϵ is referred to as dissipation rate. The higher ϵ, the faster kinetic energy is converted into heat by viscosity.

Kolmogorov's hypotheses

With Taylor's and Richardson's theories at hand, several researchers worked on a statistical description of turbulence. Among them are e.g. Karaman, Prandtl and Obukhov. Finally in 1941, Kolmogorov (1941) formulated a theory based on the results of Goldstein, Karaman, Millionshtchikov and Taylor. Kolmogorov's theory became one of the most important theories on turbulent flow. It is presented here.

It is clear that a flow field depends on its boundary conditions such as the geometry of the configuration, driving forces, etc. It is also clear that the large scales depend strongly on the configuration because the shape of the geometry will determine the large scale motions

2 Fundamentals of particle-laden flow

of the fluid.

Concerning small scale motions, the mechanisms involved are somewhat different. Kolmogorov (1941) (later translated by V. Levin, see Kolmogorov, 1991) postulated that (in the wording of Pope (2000))

> *In every turbulent flow at sufficiently high Reynolds number, the statistics of the small scale motions have a universal form that is uniquely determined by ν and ϵ.*
>
> *(Kolmogorov's first similarity hypothesis)*

Kolmogorov further introduced a length scale η_K and a time scale τ_K, based on ν and ϵ

$$\eta_K = \frac{\nu^3}{\epsilon}^{1/4}, \qquad \tau_K = \sqrt{\frac{\nu}{\epsilon}}. \tag{2.7}$$

Kolmogorov originally introduced these quantities for scaling purposes. His hypothesis means that the statistics of two turbulent flow fields are equivalent at the length and time scales η_K and τ_K. Later, η_K and τ_K were found to be the smallest length and time scales of the flow. Many authors state that this follows from Kolmogorov's hypothesis. However, Frisch (1995) shows that for this statement more unproved assumptions are necessary.

In the original paper Kolmogorov even went one step further. He postulated that

> *In every turbulent flow at sufficiently large Reynolds number, the statistics of the motions of scale $l \gg \eta_K$ are uniquely determined by ϵ and do not depend on ν.*
>
> *(Kolmogorov's second similarity hypothesis)*

As mentioned above, the large scale motions will depend on the geometry. Of course Kolmogorov was aware of that fact but he didn't mention this explicitly in his second hypothesis. Therefore Kolmogorov's second hypothesis must be restricted to scales l with $L \gg l \gg \eta_K$. Here, L denotes the largest length scale of the flow. The exact value of L depends on the configuration. For isotropic turbulence, L is usually set to the length scale of the energy containing scales, cf. section 2.1.3.

Kolmogorov's second similarity hypothesis means that, taken two turbulent flows, their statistics differ only in the large scales $l > L$. For $l < L$, the flows' statistics follow a universal law.

2.1.3 Statistical and spectral description of isotropic turbulence

Richardson and Kolmogorov considered turbulent flows in general. The present section considers one specific turbulent flow, namely homogeneous isotropic turbulent flow. This flow is characterised by

- Homogeneity: A flow is homogeneous if its statistics do not depend on the position **x**. This means that the statistics are invariant under translation of the coordinate system.

- Isotropy: A flow is isotropic if its statistics are invariant under rotation or reflection of the coordinate system.

Homogeneity means that the statistics of the flow can be described by a point value because they do not depend on **x**. Isotropy means that the statistics of $u_{f,1}(x,y,z)$ equal the statistics of $u_{f,2}(y,x,z)$ or $u_{f,3}(z,y,x)$. Therefore a homogeneous isotropic turbulent flow is the statistically simplest turbulent flow. This means that if a model for particle-laden flow shows defects in such a simple flow then it cannot be expected to perform better in a more complex configuration. This makes homogeneous isotropic turbulence well suited to the aim of this work.

Often, one speaks of 'isotropic turbulence' or 'homogeneous isotropic turbulence' instead of 'homogeneous isotropic turbulent flow'. Formally there is a difference between these expressions. For an homogeneous isotropic turbulent flow, the averaged flow field also needs to be homogeneous and isotropic whereas for homogeneous isotropic turbulence only the turbulent fluctuations need to be homogeneous and isotropic. In particular this means that the velocity of a homogeneous isotropic flow is on average zero. Thus, an isotropic turbulent flow shows isotropic turbulence but not necessarily converse.

However, concerning numerical simulation, it is common practise to speak of 'isotropic turbulence' when refering to 'homogeneous isotropic turbulent flow'. This wording is adopted hereafter.

In the present section, some results on the statistical description of isotropic turbulence are presented. Only those results are shown which are relevant for the subsequent chapters. Then, a model for the kinetic energy spectrum function E is presented. This model will be used in chapter 7. Finally, one-dimensional spectra are presented. One-dimensional spectra will be important in chapter 7 for the construction of a new model.

Statistical description of isotropic turbulence

Kolmogorov's second similarity hypothesis has consequences on the energy spectrum function E. In the following this is elaborated for isotropic turbulence.

A dimensional argument shows that in the range $\|k\| \gg 2\pi/L$ and $\|k\| \ll 2\pi/\eta_K$, the kinetic energy spectrum function must follow a power law (see e.g. Onsager, 1945; Eyink & Sreenivasan, 2006; Sreenivasan, 1995):

$$E(\|k\|) \sim \|k\|^{-5/3} \quad \text{for } 2\pi/L \ll \|k\| \ll 2\pi/\eta_K \tag{2.8}$$

For high wavenumbers, Heisenberg (1948) showed by statistical analysis that $E(\|k\|)$ must decay at least as fast as $\|k\|^{-7}$.

If one assumes that $\mathbf{u}_f$ is infinitely differentiable, then $E(\|k\|)$ must even decay faster than any polynomial (see e.g. Strichartz, 1994),

$$E(\|k\|) = O(\|k\|^{-p}) \quad \text{at } \|k\| \to \infty \text{ for any integer } p. \tag{2.9}$$

These findings give an idea on the wavenumber dependence of the energy spectrum function. Wavenumber 0 corresponds to the mean flow. For isotropic turbulence, the

2 Fundamentals of particle-laden flow

mean flow is zero and therefore the energy spectrum function must be zero at $k = 0$, $E(0) = 0$. This means that the energy spectrum function must increase at small k, show a peak at some wavenumber and then decrease again according to equations (2.8) and (2.9).

The exact location of the peak cannot be described to date but it is well understood that it is characterised by two length scales, namely the integral length scale L_f and the scale of the energy containing eddies L_k. These two scales are defined by

$$L_f = \int_{-\infty}^{\infty} f(\xi)\,\mathrm{d}\xi, \qquad f(\xi) = \frac{\overline{u_{f,1}(x,y,z)u_{f,1}(x+\xi,y,z)}}{\overline{u_{f,1}(x,y,z)u_{f,1}(x,y,z)}} \qquad (2.10\mathrm{a})$$

$$L_k = \frac{k^{3/2}}{\epsilon}. \qquad (2.10\mathrm{b})$$

f is called longitudinal two point correlation function, plotted in figure 2.1. L_f characterises the length scale over which the field decorrelates and is therefore also called (longitudinal) correlation length scale. The definition of L_f implies that a turbulent field decorrelates in space faster than ξ^{-1}. This question was already addressed by Ruelle (1986) and Egolf & Greenside (1994) but it seems that it is not fully answered. Tennekes & Lumley (1972) write that it is 'assumed that the integral scale is finite'. All experimental and numerical results support that assumption. Therefore the existence of L_f is a commonly accepted property of turbulent flows.

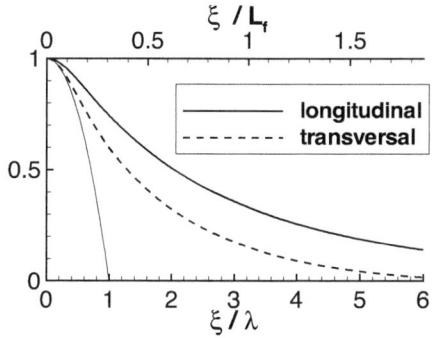

Figure 2.1: Longitudinal and transverse two point correlation functions $f(\xi)$ and $g(\xi)$, respectively. Computed from the model spectrum proposed by Pope (2000) for $Re_\lambda = 34$. The thin line shows the definition of the transverse Taylor length scale λ.

The definition of L_k is quite convenient because the ratio between L_k and the Kolmogorov length scale η_K scales with Reynolds number,

$$\frac{L_k}{\eta_K} = \frac{L_k \epsilon^{1/4}}{\nu^{3/4}} = Re_{L_k}^{3/4}, \qquad Re_{L_k} = \frac{\sqrt{k}L_k}{\nu}. \qquad (2.11)$$

With these findings it is possible to construct a so called 'model spectrum', i.e., an analytical function describing the spectrum of a homogeneous isotropic flow at a given Reynolds number, cf. section 2.1.3.

Figure 2.1 reveals an interesting fact: As a consequence of conservativity, the first derivative of the correlation functions must be zero at $\xi = 0$ (see e.g. Pope, 2000). This means that the Taylor expansion of the correlation functions around $\xi = 0$ shows no linear term. The prefactor of the quadratic term is referred to as Taylor length scale,

$$\lambda = \sqrt{-\frac{1}{2}g''(0)}, \qquad g(\xi) = \frac{\overline{u_{f,1}(x,y,z)u_{f,1}(x,y+\xi,z)}}{\overline{u_{f,1}(x,y,z)u_{f,1}(x,y,z)}}. \tag{2.12}$$

More precisely, λ defined above is the *transverse* Taylor length scale because it is computed from the *transverse* correlation function g.

One can also define a longitudinal Taylor length scale $\lambda_f = \sqrt{-\frac{1}{2}f''(0)}$. Figure 2.1 shows that $\lambda_f > \lambda$, this means that for small values of ξ the velocity decorrelates faster in transverse direction than in longitudinal direction. As a consequence of isotropy and continuity, one can derive the relation (see e.g. Pope, 2000)

$$\lambda = \frac{\lambda_f}{\sqrt{2}} \tag{2.13}$$

which leads to

$$\epsilon = 15\nu \frac{u_{rms}^2}{\lambda^2}. \tag{2.14}$$

This equation finally allows to compare the Taylor length scale against the other length scales defined so far. For example one can derive

$$\frac{L_k}{\lambda} = \sqrt{\frac{Re_{L_k}}{10}}, \qquad \frac{\lambda}{\eta_K} = \sqrt{10}Re_{L_k}^{1/4}. \tag{2.15}$$

Equation (2.14) further leads to a link between the Reynolds number defined from the large scales Re_{L_k} and the Reynolds number defined from the Taylor scale $Re_\lambda = \frac{u_{rms}\lambda}{\nu}$,

$$Re_\lambda = \sqrt{\frac{20}{3}Re_{L_k}}. \tag{2.16}$$

For more information on isotropic turbulence the reader is referred to the studies of Batchelor (1982), Tennekes & Lumley (1972), Pope (2000), Landau & Lifshitz (1987), Rotta (1972) and Hinze (1975).

Pope's model spectrum

The results from the previous section, in particular equations (2.8) and (2.9), lead to a model for the energy spectrum function E. Such a model was proposed by Kraichnan (1959). Later, Pope (2000) refined this model to read

$$E^{mod}(\kappa) = C\epsilon^{2/3}\kappa^{-5/3}f_L(\kappa L)f_\eta(\kappa\eta) \tag{2.17}$$

with the model constant $C = 1.5$. f_L and f_η are scalar functions. f_L determines the spectrum at small wavenumbers and converges to 0 for $\kappa L \to 0$ and to 1 for $\kappa L \to \infty$. f_η

2 Fundamentals of particle-laden flow

determines the spectrum at high wavenumbers (the so called dissipative range) and converges to 1 for $\kappa\eta \to 0$ and to 0 for $\kappa\eta \to \infty$. Evidently in the intermediate range, where $f_L(\kappa L) \approx f_\eta(\kappa\eta) \approx 1$, this model spectrum represents Kolmogorov's $\kappa^{-5/3}$ law, equation (2.8).

Pope specifies f_L and f_η to

$$f_L(\kappa L) = \left(\frac{\kappa L}{\left((\kappa L)^2 + c_L\right)^{1/2}} \right)^{5/3+p_0} \tag{2.18a}$$

$$f_\eta(\kappa\eta) = \exp\left(-\beta\left(\left((\kappa\eta)^4 + c_\eta^4\right)^{1/4} - c_\eta\right)\right). \tag{2.18b}$$

p_0, β, c_L and c_η are again model constants. Pope recommends $p_0 = 2$ and $\beta = 5.2$ on the basis of empirical data. The two constants c_L and c_η are set in dependence of Reynolds number such that $E^{mod}(\kappa)$ and $2\nu\kappa^2 E^{mod}(\kappa)$ integrate to the flow's kinetic energy k and turbulent dissipation rate ϵ, respectively. The model spectrum is shown in figure 2.2 for $Re_\lambda = 99$ and $Re_\lambda = 265$.

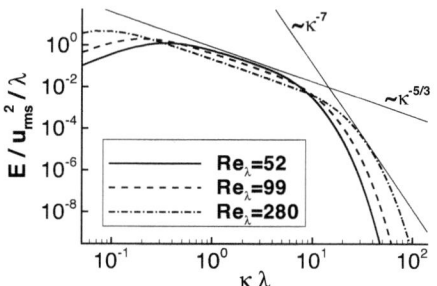

Figure 2.2: Model spectrum propsed by Pope (2000) for $Re_\lambda = 52$, $Re_\lambda = 99$ and $Re_\lambda = 280$ together with lines proportional to $\kappa^{-5/3}$ and κ^{-7}.

One-dimensional spectra

So far, the statistical description of the flow field was based on the energy spectrum function E. Going back to the definition of E, equation (2.4), one can see that $E(\kappa)$ is an average in spectral space over wavenumbers $\mathbf{k}'$ of uniform length, $\|\mathbf{k}'\| = \kappa$. The idea behind this is that on average the norm of the Fourier transform is a spherical function,

$$\overline{\|\mathcal{FT}(\mathbf{u}_f)(\mathbf{k})\|^2} = \frac{E(\|\mathbf{k}\|)}{4\pi\|\mathbf{k}\|^2}. \tag{2.19}$$

In isotropic turbulence this is true, but only as long as the Fourier transform of all three velocity components $\mathcal{FT}(\mathbf{u}_f)$ is concerned. If one considers the Fourier transform of individual components $\mathcal{FT}(u_{f,i})$, then one will find that these functions are not spherical but that they are different in direction of k_i and k_j, $j \neq i$. More precisely, for each velocity component $u_{f,i}$ one must distinguish between the respective longitudinal wavenumber k_i and the transverse wavenumbers k_j, $j \neq i$. Because of isotropy, the transverse wavenumbers are interchangeable, for example

$$\overline{|\mathcal{FT}(u_{f,1})(k_1,k_2,k_3)|^2} = \overline{|\mathcal{FT}(u_{f,1})(k_1,k_3,k_2)|^2} = \overline{|\mathcal{FT}(u_{f,2})(k_2,k_1,k_3)|^2}. \qquad (2.20)$$

On the other hand, transverse and longitudinal wavenumbers are not to be confused,

$$\overline{|\mathcal{FT}(u_{f,1})(k_1,k_2,k_3)|^2} \neq \overline{|\mathcal{FT}(u_{f,1})(k_2,k_1,k_3)|^2}. \qquad (2.21)$$

In experiments it is easier to measure only one velocity component along one dimension instead of all three components in three dimensions. This inspires to use one-dimensional spectra $E_{j_1 j_2}$, the one-dimensional Fourier transform of a single velocity component, averaged in the remaining two directions,

$$E_{j_1 j_2}(\kappa) = \int_{\mathbf{R}^3} \left| u_{j_1,f}(\mathbf{x}) e^{-i\kappa x_{j_2}} \right|^2 \, \mathrm{d}\mathbf{x}. \qquad (2.22)$$

Due to Parseval's theorem, $E_{j_1 j_2}$ can also be expressed using the three-dimensional Fourier transform $\mathcal{FT}$ and the Dirac delta function δ,

$$E_{j_1 j_2}(\kappa) = \int_{\mathbf{R}^3} \delta_{k_{j_2}}(\kappa) \left| \mathcal{FT}(u_{j_1,f})(\mathbf{k}) \right|^2 \, \mathrm{d}\mathbf{k}. \qquad (2.23)$$

For example, E_{12} reads

$$E_{12}(\kappa) = \int_{\mathbf{R}^2} \left| \mathcal{FT}(u_{1,f}) \begin{pmatrix} k_1 \\ \kappa \\ k_3 \end{pmatrix} \right|^2 \, \mathrm{d}k_1 \, \mathrm{d}k_3. \qquad (2.24)$$

In isotropic turbulence, E_{11}, E_{22} and E_{33} are statistically identical. These spectra are called *longitudinal spectra* E_l,

$$E_l = E_{11} = E_{22} = E_{33}. \qquad (2.25)$$

Also the remaining spectra are identical, called *transverse spectra* E_t,

$$E_t = E_{12} = E_{13} = E_{21} = E_{23} = E_{31} = E_{32}. \qquad (2.26)$$

On first sight it seems that the ensemble of all one-dimensional spectra provides more information than the energy spectrum function E. In isotropic turbulence this is not true. For example Pope (2000) shows how to compute E_l and E_t from E and vice versa.

2.2 Particle-laden turbulent flow

The previous section focused on single phase flow, in particular on isotropic turbulence. In the present chapter, particle-laden flow is discussed.

Actually 'particle-laden flow' covers a wide range of two phase flows. The present section contains a classification of particle-laden flows and a specification of the class under consideration in this work. The mathematical description of choice for the class under consideration is the so called Euler–Lagrange approach or, more precisely, point-particle Euler–Lagrange approach. This approach needs an equation of motion for an individual particle, the Maxey–Riley equation.

In the following, first a classification of particle-laden flows is presented. The terms 'point-particle' and 'Euler–Lagrange' are explained. Then, a section on the Maxey–Riley equation and its history follows. Finally, physical effects are discussed which can be observed in isotropic turbulence with the class of particles under consideration.

The presented results are mainly based on the studies of Berlemont *et al.* (1990), Maxey & Riley (1983), Taylor (1953, 1954), Maxey (1987), Fessler *et al.* (1994), Clift *et al.* (1978) and Crowe *et al.* (1998).

2.2.1 Classification and specification

In the introduction (chapter 1), two examples for particle-laden flow were given. One was sediment transport and the other distribution of aerosols in the atmosphere. These examples give an idea of what is meant by 'particle-laden flow' in the context of this work.

More precisely, only

- rigid,
- spherical,
- non-rotating particles

are considered. These particles can be further classifed with respect to

- particle diametre,
- material density
- and the number of particles in the flow.

The latter three parameters determine which mathematical description of the particles is appropriate as well as requirements for coupling of carrier flow and particles. Both questions are addressed in the present section. At the end of the section, a precise specification of the class under consideration in the present work is given.

Classification with respect to mathematical modelling

There are various approaches for the mathematical description of particles immersed in a fluid and none of them is suited for every possible application. In view of a numerical computation of particle-laden flow, the mathematical description must be such that, on one hand the computational requirements can be fulfilled and on the other hand the accuracy of the mathematical description is as high as possible. The particle diametre is a limit for the accuracy of the description, the number of particles is limited by computational possibilities. Table 2.1 gives an overview on the possible mathematical descriptions that are dependent on these two parameters. The listed approaches are described below. The remaining work

Table 2.1: Classification of particle-laden flow in view of numerical simulation with respect to an appropriate mathematical description. The mathematical description with the highest accuracy is listed for each class. d denotes particle diametre, η_K is the Kolmogorov length scale of the carrier flow and N_p the number of particles per CPU. Limiting N_p for resolved particle approach is based on the work of Uhlmann (2008), limiting N_p for Lagrangian approach is based on own computations (one-way coupling, cf. section 2.2.1).

d \ N_p	$\lesssim 10$	$\lesssim 10^6$	$> 10^6$
$< \eta_K$	Euler–Lagrange point-particle or resolved particle	Euler–Lagrange point-particle	Euler–Euler
$\geq \eta_K$	resolved particle	n.a.	n.a.

focuses on the Euler–Lagrange approach. More information on up to date developments for each approach can be found in Balachandar & Eaton (2010).

Table 2.1 is based on state of the art computational methods and todays computers. Even with the capabilities of todays supercomputers, there does not exist any mathematical description which allows numerical simulation if the number of particles is high and the particles are larger than the Kolmogorov length scale (cf. e.g. Balachandar & Eaton, 2010).

The three approaches mentioned in table 2.1 can be explained as follows:

- *Resolved particle approach.* The resolved particle approach is the approach with the least simplifications (and thus with the most details). Here, each particle is treated as a solid body immersed in the fluid. Similar to for fluid-structure interaction problems, exact equations can be derived for the force of the fluid acting on the particle and vice versa. More precisely, the equations describe the force of the fluid on each point on the particle's surface. However, this approach is limited to a relatively small number of particles. Recently, Uhlmann (2008) conducted a simulation with 4096 particles. This simulation was conducted on 512 CPUs and required approximately 10^6 CPU hours. Most applications involve far more particles and therefore in the near future the resolved particle approach will remain a subject of research.

- *Euler–Euler approach.* Somewhat to the other extreme is the treatment of all suspended particles as one continuous phase that is transported with the flow. This means that the particles are described by spatio-temporal fields such as concentration $c(\mathbf{x}, t)$ and velocity $\mathbf{u}_p(\mathbf{x}, t)$. This approach is referred to as Euler–Euler approach because the carrier flow and the suspended phase are treated in a Eulerian context, i.e., by discretisation in space and time. $\mathbf{u}_p(\mathbf{x}, t)$ denotes the average particle velocity within the cell assigned to $\mathbf{x}$. With an Euler–Euler approach one cannot derive governing equations for $\mathbf{u}_p(\mathbf{x}, t)$ from first principles. Modelling and uncertainties come into play but, on the other hand, computational requirements can be held relatively low. Up to date studies on Euler–Euler approaches can be found in Simonin *et al.*

2 Fundamentals of particle-laden flow

(2006) or Shotorban & Balachandar (2007). However, also their models are restricted to particles which are smaller than the Kolmogorov length scale.

- *Euler–Lagrange point-particle approach.* A mathematical description that lies between the resolved particle and Euler–Euler approach is treatment of the particles as point particles. Hereafter, this is referred to as Euler–Lagrange approach. The carrier flow is solved in a Eulerian context, the particles in Lagrangian context, the suspended phase is discretised by single particles. 'Point particles' does not mean that they are infinitesimally small and massless but it means that the particles are smaller than the smallest scales of the flow. This allows modelling the effect of the fluid on the particle through a single point force, located at the centre of the particle. The present work follows this approach. The constitutive equation for the particles is defined in section 2.2.2.

Classification with respect to coupling of fluid flow and particles

Independent of the mathematical description for the particles, one must take into consideration that particles and fluid have an effect on each other. Therefore governing equations for the fluid and governing equations for the particles must be coupled. Table 2.2 gives an overview on the type of effects.

Table 2.2: Classification of particle-laden flow with respect to equation coupling following Elghobashi (1991). Φ denotes the volume fraction.

Φ	effect(s)	coupling approach
$< 10^{-6}$	fluid $\rightarrow$ particle	one-way coupling
10^{-6} to 10^{-3}	fluid $\rightarrow$ particle and particle $\rightarrow$ fluid	two-way coupling
10^{-3} to 1	fluid $\rightarrow$ particle and particle $\rightarrow$ fluid and particle $\leftrightarrow$ particle	four-way coupling
> 1	collision dominated	n.a.

In detail, the coupling approaches and corresponding effects can be explained as follows.

- *One-way coupling.* The carrier fluid has an effect on the particles but the effect of particles on the carrier fluid or inter-particle effects are negligible. This is valid for dilute suspensions.

- *Two-way coupling.* The carrier fluid has an effect on the particles (1st way) and the particles have an effect on the fluid (2nd way) but inter-particle effects are negligible. This is valid for moderately dense suspensions.

- *Four-way coupling.* The carrier fluid has an effect on the particles (1st way) and the particles have an effect on the fluid (2nd way). In addition, the particles have an effect onto each other, because the proximity of one particle affects the surrounding fluid of another particle (3rd way) and because particles may collide (4th way). This is valid for dense suspensions.

- *Collision dominated regime.* For volume fractions higher than 1, the particle's dynamics become collision dominated. This regime is called granular flow. Such flows must be treated with very different methods that are beyond the scope of this work.

In table 2.2, the magnitude and significance of the respective effects is rated in dependence of the particle volume fraction, i.e., the ratio between the volume occupied by particles and the volume occupied by particles and fluid. These rules of thumb were given by Elgobashi already in 1991. Later, Sundaram & Collins (1999) showed that the mass fraction must be taken into consideration as well but they did not give any explicit numbers on volume or mass fraction. In recent studies (see e.g. Vreman *et al.*, 2009; Apte *et al.*, 2008) it is common practise to mention the mass fraction and classify the flow according to the volume fraction as proposed by Elghobashi (1991).

Concerning moderately dense suspensions where two-way coupling is an issue, there are still open problems. Balachandar & Eaton (2010) point out that state of the art resolved particle methods are capable to predict such flows correctly but that Euler–Lagrange methods do not. On the other hand, for dilute suspensions where one-way coupling is admissible, state of the art methods produce reliable results.

The present work does not concentrate on coupling issues. Therefore the present work considers only dilute suspensions and follows a one-way coupling approach.

Specification of the class of particle-laden flow under consideration

The present work considers particle-laden flows with the following properties:

1. the particles are rigid and spherical
2. the particles are smaller than the Kolmogorov length scale
3. the particles are non-rotating
4. the volume fraction is smaller than 10^{-6}
5. the ratio between the material density of the particles and the material density of the carrier fluid is higher than 1000
6. the number of particles is small enough such that computational ressources allow treatment in an Euler–Lagrange context.

Property 4 allows for one-way coupling. In some numerical simulations, the volume fraction Φ will be higher than 10^{-6} in order to obtain more statistical samples. The results from these simulations must be interpreted as the statistical average of N simulations of one particle-laden flow. N is such that the volume fraction Φ/N in this flow is smaller than 10^{-6}.

Property 5 will be of importance concerning the governing equations for the particles, cf. section 2.2.2. Property 6 evidently means that in the numerical simulations up to 10^6 particles can be traced per CPU.

2.2.2 Maxey–Riley equation and its history

As mentioned above, the present work follows a Euler–Lagrange approach with one-way coupling. With this approach, multiple particles are immersed in a carrier fluid and each particle can be treated independently of the others.

In addition to the Navier–Stokes equations, the approach needs only one more constitutive equation, namely an equation of motion for a single immersed particle. The equation must describe the effect of the flow on the particle.

This effect depends strongly on the particle Reynolds number Re_p, defined by

$$Re_p = \frac{d\|\mathbf{u}_p - \mathbf{u}_{f@p}\|}{\nu} \tag{2.27}$$

Here, d denotes the particle diametre and $\mathbf{u}_p$ is the velocity of the particle. $\mathbf{u}_{f@p}$ is the velocity of the carrier fluid at the absence of the particle and is called *fluid velocity seen by the particle*. Re_p characterises the complexity of the flow in the immediate surrounding of the particle. Low Re_p means low complexity, high Re_p means high complexity.

Even under the assumption that the particle Reynolds number Re_p is very small, it took almost one century to derive an equation of motion for a single immersed particle. This equation, referred to as Maxey–Riley equation, is presented in the present section. If the Reynolds number is high, then the state of the art is to include empirical corrections into the Maxey–Riley equation.

In the following first a brief historical review on the derivation of the Maxey–Riley equation is given and then corrections for high Reynolds number are presented. Finally, the equation is simplified for high ratios between the material density of the particles and the material density of the carrier fluid.

Aiming at an exact equation of motion for a single particle

More than a century ago, Basset (1888) analysed a settling particle in a fluid that was otherwise at rest. He derived an equation of motion for the particle under the assumption that the particle Reynolds number is suffiently small such that the flow around the particle is a Stokes flow. Later, Boussinesq (1903) obtained the same equation independently. His work was continued by Oseen (1927), leading to the Basset-Boussinesq-Oseen equation which is valid for uniform flow at small Re_p.

In order to obtain an equation that is valid for non-uniform flows, many authors refined the Basset-Boussinesq-Oseen equation. Among them are Tchen (1947), Corrsin & Lumley (1956), Buevich (1966) and Riley (1971), just to mention a few. Tchen aimed at an equation valid for an unsteady non-uniform carrier flow. His extension showed promising results but included some unjustified assumptions. It reads (the physical meaning of the individual terms is discussed below)

$$\begin{aligned}\frac{\mathrm{d}u_{p,i}}{\mathrm{d}t} &= \frac{\rho_p - \rho_f}{\rho_p}g_i + \frac{\rho_f}{\rho_p}\left(\frac{\partial u_{f,i}}{\partial t} + u_{p,j}\frac{\partial u_{f,i}}{\partial x_j}\right) + \frac{1}{\tau_p}\left(u_{f@p,i} - u_{p,i}\right) \\ &+ \frac{\rho_f}{2\rho_p}\frac{\mathrm{d}}{\mathrm{d}t}\left(u_{f@p,i} - u_{p,i}\right) + \frac{\mathrm{d}}{2\tau_p}\int_0^t \frac{\frac{\mathrm{d}}{\mathrm{d}t}\left(u_{f@p,i} - u_{p,i}\right)}{\sqrt{\pi\nu(t-\tau)}}\mathrm{d}\tau.\end{aligned} \quad (2.28)$$

The notations stand for

- $\mathbf{u}_p(t)$: particle velocity,
- ρ_f and ρ_p: material density of carrier fluid and suspended particles, respectively,
- $\mathbf{g}$: gravity,
- $\mathbf{u}_{f@p}(t) = \mathbf{u}_f(t, \mathbf{x}_p(t))$: velocity of the carrier fluid at the absence of the particle, called *fluid velocity seen by the particle*,
- $\mathbf{x}_p(t)$: particle position,
- $\frac{\partial u_{f,i}}{\partial t} + u_{p,j}\frac{\partial u_{f,i}}{\partial x_j}$: material derivative of the velocity of the carrier fluid following the moving particle, evaluated at $(t, \mathbf{x}_p(t))$,
- $\tau_p = \frac{\rho_p}{\rho}\frac{d^2}{18\nu}$: particle relaxation time,
- d: particle diametre,
- ν: kinematic vicosity of the carrier fluid.

Tchens equation showed several deficiencies and was heavily discussed by Corrsin & Lumley (1956), Soo (1975) and Gitterman & Steinberg (1980). In 1983, almost a century after Basset's work, Maxey & Riley (1983) gave an answer to these discussions. They derived an equation of motion for particles from first principles and arrived at an equation which is different to Tchen's equation but coincides with the latter for uniform flow. It reads

$$\frac{\mathrm{d}u_{p,i}}{\mathrm{d}t} = \frac{\rho_p - \rho_f}{\rho_p}g_i \quad (2.29\mathrm{a})$$

$$+ \frac{\rho_f}{\rho_p}\left(\frac{\partial u_{f,i}}{\partial t} + u_{f@p,j}\frac{\partial u_{f,i}}{\partial x_j}\right) \quad (2.29\mathrm{b})$$

$$+ \frac{1}{\tau_p}\left(u_{f@p,i} - u_{p,i} + \frac{d^2}{24}\frac{\partial^2 u_{f,i}}{\partial x_j^2}\right) \quad (2.29\mathrm{c})$$

$$+ \frac{\rho_f}{2\rho_p}\frac{\mathrm{d}}{\mathrm{d}t}\left(u_{f@p,i} + \frac{d^2}{40}\frac{\partial^2 u_{f,i}}{\partial x_j^2} - u_{p,i}\right) \quad (2.29\mathrm{d})$$

$$+ \frac{\mathrm{d}}{2\tau_p}\int_0^t \frac{\frac{\mathrm{d}}{\mathrm{d}t}\left(u_{f@p,i} - u_{p,i} + \frac{d^2}{24}\frac{\partial^2 u_{f,i}}{\partial x_j^2}\right)}{\sqrt{\pi\nu(t-\tau)}}\mathrm{d}\tau. \quad (2.29\mathrm{e})$$

2 Fundamentals of particle-laden flow

Terms (2.29a) and (2.29b) are called buoyancy and fluid acceleration force. The fluid acceleration force incorporates the effect of the pressure gradient and viscous drag on the particle's surface, derived by employing the Navier-Stokes equation with a no-slip condition on the particle's surface. These two terms are the only forces acting on the particle if the effect of the particle on the surrounding fluid is neglected (see Riley, 1971). Of course, this is only valid if $d = 0$ which is not very interesting.

Maxey & Riley (1983) considered the effect of the particle on the surrounding fluid as well. Then, the remaining terms of equation (2.29) come into play. These are grouped with respect to the time derivatives. Term (2.29c) is the viscous Stokes drag (also referred to as aerodynamical drag), mainly resulting from the slip velocity between particle and fluid $\|u_{f@p} - u_p\|$. One often refers to $\frac{d^2}{24\tau_p} \frac{\partial^2 u_{f,i}}{\partial x_j^2}$ as Faxen correction. Faxen, Wiman & Oseen (1922) proposed this term as correction term for the Stokes drag on a particle immersed in a pipe flow at small particle Reynolds numbers. This term actually stands for streamline curvature effects (see Maxey & Riley, 1983). (2.29d) and (2.29e) are called added mass and Basset history term. Added mass can be interpreted as the force which occurs because the particle cannot move isolated through the fluid but must always move the surrounding fluid as well. Finally, the structure of the surrounding flow and therefore also the drag force depends on the history of the particle. This is covered by the Basset history term. Therefore this term can be interpreted as correction for the Stokes drag.

Extensions for high particle Reynolds numbers

Maxey and Riley derived their equation for small particle Reynolds numbers. They needed this assumption in order to describe the modification of the flow induced by the presence of the sphere. At high Reynolds number, this modification can be very complex.

Taneda (1956) analysed experimentally the flow behind a sphere at various Reynolds numbers. He dragged a steel ball through water at various velocities and obtained the flow fields depicted in figure 2.3. At low Reynolds number, the flow simply passes the sphere, no separation occurs. This is the case which Maxey and Riley analysed. At increasing Reynolds number (around 20), the flow separates from the sphere, wakes and recirculation zones evolve. More detailed descriptions of these processes can be found in the original paper by Taneda (1956) or in the book of Clift, Grace & Weber (1978).

The state of the art is to respect such effects by including empirical correction coefficients in the equation derived by Maxey and Riley. One defines three correction coefficients, namely one for Stokes drag (2.29c), one for added mass (2.29d) and one for the Basset history term (2.29e) (see Berlemont et al., 1990). Terms (2.29a) and (2.29b) stem directly from the Navier-Stokes equation. They are independent of Re_p and do not need any correction coefficients.

In the following, only Stokes drag (2.29c) will be considered (see below). In order to include corrections for arbitrary particle Reynolds number, (2.29c) is replaced by

$$\frac{c_D Re_p}{24\tau_p} \left(u_{f@p,i} - u_{p,i} + \frac{d^2}{24} \frac{\partial^2 u_{f,i}}{\partial x_j^2} \right) \tag{2.30}$$

where the drag coefficient c_D depends on particle Reynolds number Re_p due to separation, recirculation zones and vortex shedding, cf. figure 2.3. With this modification, Stokes drag

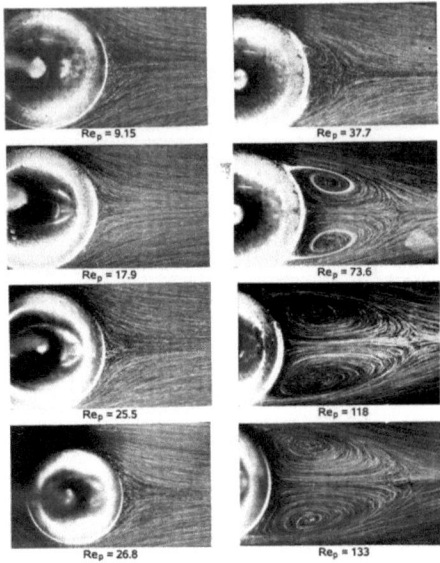

Figure 2.3: Flow behind a sphere. Photographs taken by Taneda (1956) (reprinted with permission).

becomes a non-linear term.

Clift et al. (1978) recommend a piecewise defined function for $c_D(Re_p)$, based on 28 studies of various authors. This function is shown in figure 2.4 together with the widely used approximation proposed by Schiller & Naumann (1933)

$$c_D = \frac{24}{Re_p}\left(1 + 0.15 Re_p^{0.687}\right) \qquad \text{approximation proposed by Schiller \& Naumann (1933).}$$
(2.31)

Figure 2.4 shows that Schiller and Naumann's curve is recommendable up to $Re_p \approx 800$. The region beyond $Re_p = 800$ can be divded into a 'high subcritical' range ($400 < Re_p < 3.5 \times 10^5$) and a 'supercritical' ($Re_p > 3.5 \times 10^5$) range (see Clift et al., 1978).

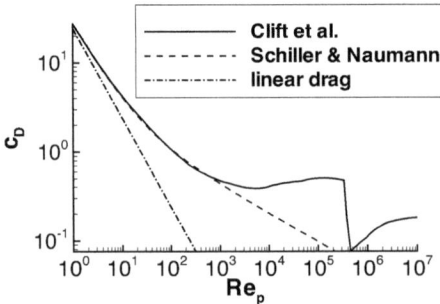

Figure 2.4: Drag coefficient of a sphere as a function of particle Reynolds number according to Clift et al. (1978) and Schiller & Naumann (1933). For linear drag, $c_D = 24/Re_p$ holds.

In the high subcritical range, a three-dimensional rotating wake evolves behind the sphere (see Seeley et al., 1975) and periodical vortex shedding occurs at the rear surface (see Achenbach, 1974). Directly behind the sphere, the wake is laminar, further downstream a turbulent wake evolves (see Brennen, 2005). It seems to be unclear whether in this range the flow behind the sphere should be considered turbulent. Clift et al. (1978) explicitly point out that it should not, in accordance with the laminar wake near the sphere. On the other hand, the turbulent wake behind the sphere affects the drag coefficient c_D and therefore the flow can be considered turbulent (see Brennen, 2005). However, concerning the boundary layer, there is no ambiguity. Below $Re_p = 3.5 \times 10^5$, it is laminar and around $Re_p = 3.5 \times 10^5$ the boundary layer becomes turbulent, resulting in a drop of the drag coefficient (see Clift et al., 1978; Achenbach, 1974).

In the simulations presented in this work, the particle Reynolds number computed from average slip velocity was quite small (only up to $Re_p = 10$). However, this does not neccessarily mean that Schiller and Naumann's approximation is admissible because, as shown for example by Toschi & Bodenschatz (2009) or Biferale et al. (2004), high peak values can occur. The latter authors analysed a configuration which is comparable to the configurations analysed in the present work and found that the particle slip velocity can easily attain 80 times of the rms value. Then, the peak particle Reynolds number is $Re_p = 800$, i.e., Re_p attains the limit of admissibility for Schiller and Naumann's approximation. In order to achieve higher accuracy, Clift et al.'s recommendation was implemented in the present work.

Simplifications

The equation proposed by Maxey and Riley is rarely solved in the form stated above. The practical reason is simply that in this form the computational requirements are so high that one would be limited to a very small number of particles.

However, if the material density of the particles ρ_p is very much higher than the density of the carrier fluid ρ_f, then the prefactors for the fluid acceleration force (2.29b) and the added mass term (2.29d) are small. Although there is no rigorous justification, it is generally accepted that fluid acceleration force and added mass are negligible if ρ_p/ρ_f is higher than 1000 (cf. Nguyen & Schulze, 2003; Fukagata, 1998). A similar and generally accepted argument states that for small particles the Basset history term is negligible (cf. Nguyen & Schulze, 2003) and that streamline curvature effects are negligible as long as the focus is not on near wall effects (cf. Mittal, 1993)

As mentioned above, there is no rigorous justification for these simplifications but the numerical experiments of Armenio & Fiorotto (2001) and Kubik & Kleiser (2004) are in accordance with them. Therefore also in very recent studies (e.g. Guha, 2008; Vreman, 2007; Almeida & Jaberi, 2008; IJzermans, Hagmeijer & van Langen, 2007; Gui, Fan & Cen, 2008) the authors rely on these arguments and solve

$$\frac{du_{p,i}}{dt} = \frac{c_D Re_p}{24\tau_p}\left(u_{f@p,i} - u_{p,i}\right) \qquad \text{(simplified Maxey–Riley equation)}. \qquad (2.32)$$

Hereafter, equation (2.32) is referred to as 'simplified Maxey–Riley equation'. The present work follows this approach. It turned out that with the computational resources available, approximately one million particles per CPU core can be traced simultaneously.

With equation (2.32), the suspended particles can be described by two dimension-free parameters, namely particle Reynolds number Re_p and Stokes number St

$$Re_p = \frac{d\|\mathbf{u}_p - \mathbf{u}_{f@p}\|}{\nu}, \qquad St = \frac{\tau_p}{\tau_K}. \tag{2.33}$$

Here, τ_K denotes the Kolmogorov time scale (cf. section 2.1.2). For small particle Reynolds numbers, Stokes drag is linear and equation (2.32) becomes

$$\frac{du_{p,i}}{dt} = \frac{1}{\tau_p}\left(u_{f@p,i} - u_{p,i}\right). \tag{2.34}$$

This equation can be solved analytically,

$$\mathbf{u}_p(t) = \mathbf{u}_p(t_0)e^{\frac{t_0-t}{\tau_p}} + \frac{1}{\tau_p}\int_{t_0}^{t}\mathbf{u}_{f@p}(\tau)e^{\frac{\tau-t}{\tau_p}}\,d\tau. \tag{2.35}$$

It should be noted that this equation is implicit in the particle velocity $\mathbf{u}_p$ because $\mathbf{u}_{f@p}$ must be evaluated along the particle's path, i.e., along the path determined by $\mathbf{u}_p$. In the present work this equation is only used for analytical assessment of particle-LES models, sections 6.3 and 7.3.

2.2.3 Turbulent dispersion and preferential concentration

As mentioned above, the present work considers point particles, governed by the Maxey–Riley equation. If such particles are immersed in isotropic turbulence, then one can observe two important effects, namely turbulent dispersion and preferential concentration. These two effects are explained in the following.

Turbulent dispersion

Already in 1953, Taylor (1953, 1954) studied particles immersed in a turbulent flow. In that work, he considered inertia free particles in isotropic turbulence but his results are valid for inert particles as well.

A cloud of particles immersed in a turbulent flow will always grow in size. This effect can be observed for example at the smoke leaving a chimney and is called *turbulent dispersion*. Turbulent dispersion means that the average distance of two particles in the cloud increases with time.

In isotropic turbulence, dispersion can be quantified by measuring the distance of a particle from the point of initialisation $\|\mathbf{x}_p(t) - \mathbf{x}_p(0)\|$. The faster this distance increases, the higher is the rate of dispersion D.

Taylor (1953, 1954) formulated the rate of dispersion D in terms of the kinetic energy of

2 Fundamentals of particle-laden flow

the particles k_p and a time scale t_p,

$$D = \lim_{t \to \infty} \frac{\mathrm{d}\overline{\|\mathbf{x}_p(t) - \mathbf{x}_p(0)\|^2}}{\mathrm{d}t} = 2 \lim_{t \to \infty} \int_0^t \overline{u_{p,i}(t)u_{p,i}(\tau)} \, \mathrm{d}\tau = 4 k_p t_p \qquad (2.36\mathrm{a})$$

$$\text{with} \quad k_p = \frac{1}{2}\overline{u_{p,i}^2} \quad \text{and} \quad t_p = \lim_{t \to \infty} \int_0^t \frac{\overline{u_{p,i}(t)u_{p,i}(\tau)}}{\overline{u_{p,i}(t)u_{p,i}(t)}} \, \mathrm{d}\tau. \qquad (2.36\mathrm{b})$$

$\overline{\cdot}$ denotes averaging over particles. t_p is called correlation or integral time scale. For turbulent flow, t_p is finite for $t \to \infty$ because the velocity seen by a particle $\mathbf{u}_{f@p}$ decorrelates.

The notation above is valid for statistically steady particle dynamics. Otherwise, the time dependence of k_p must be taken into consideration additionally.

Concluding, Taylor's result was that turbulent dispersion can be quantified via kinetic energy and integral time scale.

Preferential concentration

Taylor considered a particle cloud as a whole. If one takes a closer look into the cloud then one will observe another effect, namely *preferential concentration*. Preferential concentration stands for particle clustering due to the interaction of centrifugal forces and Stokes drag. It can be observed in homogeneous and in inhomogeneous turbulent flow.

Actually preferential concentration was first observed without noticing it. Sehmel (1980) observed that the sink velocity of a particle due to gravity is higher in a turbulent flow than in the quiescent fluid. He was not aware that this is because of an inhomogeneous particle distribution. Maxey (1987) analysed this effect by theoretical considerations and kinematic simulation. He defined a continuous Eulerian particle velocity field, obtained by averaging over the particles, and showed that the sink velocity depends on the divergence of the particle velocity field. Then, he showed that the divergence depends on the vorticity and strain rate of the carrier flow and concluded that particles cluster in regions of low vorticity and high strain rate. Thus, it was Maxey who explained that Sehmel's observations were due to particle clustering.

Later, Squires & Eaton (1990, 1991) and Wang & Maxey (1993) provided DNS data at low Reynolds number supporting Maxey's theory. They showed that, in dependence of Stokes number, particles may not be distributed homogeneously even if the flow is homogeneous. Squires & Eaton (1990) named this effect as 'preferential concentration'. Figures 2.5 and 2.6 give an idea of the effect. The figures show that the intensity of preferential concentration depends on Stokes number.

This dependency was quantified by Fessler, Kulick & Eaton (1994). They provided data from an experiment of particle-laden turbulent channel flow. Fessler *et al.* (1994) introduced a measure for preferential concentration by dividing the test section into boxes and computing a histogram of the number of particles per box. This histogram leads to the distribution P_{pc}. For homogeneously distributed particles, P_{pc} is the Poisson distribution $P_{Poisson}$. If preferential concentration occurs then P_{pc} and $P_{Poisson}$ differ. The first moment of these distributions is the average number of particles per box and therefore the first moments of

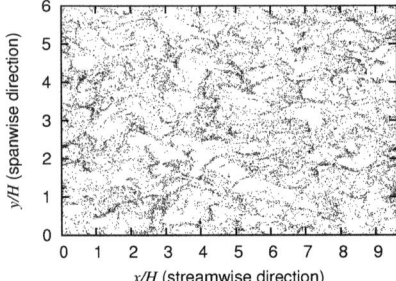
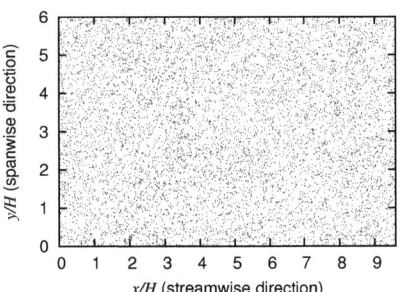

Figure 2.5: DNS of Lycopodium in turbulent channel flow (air), instantaneous distribution on channel centreplane, $Re_\tau = 180$, $St = 0.6$. H denotes channel half height. (for details cf. Gobert et al., 2007)

Figure 2.6: DNS of copper particles in turbulent channel flow (air), instantaneous distribution on channel centreplane, $Re_\tau = 180$, $St = 56$. H denotes channel half height. (for details cf. Gobert et al., 2007)

both distributions are equal. Thus, one compares second moments and defines as measure for preferential concentration

$$\Sigma = \frac{\sigma_{pc} - \sigma_{Poisson}}{\sigma_{Poisson}^2}. \qquad (2.37)$$

Here, σ denotes the standard deviation of P_{pc} and $\sigma_{Poisson}$ denotes the standard deviation of $P_{Poisson}$. For the Poisson distribution, the first moment equals $\sigma_{Poisson}^2$, therefore it makes sense to normalise Σ by $\sigma_{Poisson}^2$.

Fessler et al. (1994) found that preferential concentration is strongest around $St = 1$, i.e., Σ is maximal when particle relaxation time equals Kolmogorov time. Their results were later supported by DNS and LES of turbulent channel flow and isotropic turbulence at various Reynolds number (cf. Rouson & Eaton, 2001; Reade & Collins, 2000; Hogan & Cuzzi, 2001; Bec et al., 2007; Wang & Squires, 1996; Yang & Lei, 1998).

A somewhat open question is the dependence of preferential concentration on Reynolds number. Several authors attacked this problem by DNS of isotropic turbulence but obtained somewhat contradictory results. Hogan & Cuzzi (2001) could not find a difference in preferential concentration between $Re_\lambda = 40$, $Re_\lambda = 80$ and $Re_\lambda = 140$. On the other hand, Reade & Collins (2000) and Wang et al. (2000) found that at $Re_\lambda = 82.5$ preferential concentration is stronger than at $Re_\lambda = 30$. Wang et al. even predict linear increase of preferential concentration with Reynolds number, i.e., no saturation occurs. On the other hand, Collins & Keswani (2004) conducted DNS of isotropic turbulence up to $Re_\lambda = 152$ and found sublinear behaviour. They predict that preferential concentration saturates with Reynolds number, in contradiction with the results of Wang et al. (2000). This issue is not yet clarified.

It is well known that preferential concentration does not lead to clustering in the sense

2 Fundamentals of particle-laden flow

that the particles would form compact balls. Figures 2.5 and 2.6 show that the particles rather tend to align on a surface. All of the previously mentioned authors made the same observation. Therefore Rouson & Eaton (2001) proposed an alternative measure to Σ, namely the fractal dimension of the suspended phase. The idea is that if the particles are aligned on surfaces then they form a two-dimensional shape. This means that if one centres a ball with radius r on some particle and counts the number N of particles which reside within this ball, then on average N scales with $N \sim r^2$ as long as the curvature of the surface within the ball is negligible. On the other hand, if the particles are distributed homogeneously then they form a three-dimensional shape, $N \sim r^3$. Preferential concentration actually leads to something in between. One assigns to that shape a non-integer fractal dimension d_{pc}, defined by

$$N \sim r^{d_{pc}}, \tag{2.38}$$

where r covers a range of radii. r but must be sufficiently small such that d_{pc} is approximately constant. At the same time, r must be large enough such that a sufficiently large number of particles resides within the ball. Grassberger & Procaccia (1983) first introduced this measure in the context of dynamical systems in their (nicely titled) work 'Measuring the Strangeness of Strange Attractors'. Tang et al. (1992) were the first to apply this measure in the context of particle-laden flow. Concerning preferential concentration, d_{pc} equals 3 for $St = 0$ and $St \to \infty$ and in all studies on preferential concentration it was observed that d_{pc} shows a minimum around $St = 1$ in accordance with the behaviour of Σ.

It should be noted that preferential concentration can only be detected if the number density of the particles is sufficiently high. In this case, coupling issues come into play (cf. section 2.2.1). In particular, Geurts & Vreman (2006) and Vreman et al. (2009) showed that collisions can significantly affect preferential concentration.

As mentioned above, the present thesis follows a one-way coupling approach. Therefore the results presented here are only valid for sufficiently small volume fractions. In particular, the results on preferential concentration are only valid for particles with high material density and small diametre.

3 Implemented simulation tools

> *There are 3 rules to follow when parallelizing large codes. Unfortunately, no one knows what these rules are.*
> W. Somerset Maugham as paraphrased by Gary Montry

This chapter covers the numerical tools which were implemented in the present work. Section 3.1 discusses two different simulation techniques for turbulent flow, namely Direct Numerical Simulation and Large Eddy Simulation. Section 3.2 covers numerical methods used for computing the carrier flow. This includes a brief description of the discretisation schemes and a description of the applied forcing scheme. A forcing scheme is a method to drive isotropic turbulence, i.e., to compensate for dissipation. Section 3.3 gives details on the numerical methods used for the computation of the particles. Section 3.4 contains a list of all implemented simulation techniques and numerical methods.

3.1 Simulation techniques for turbulent flow

This chapter covers two techniques for simulation of turbulent single phase flow, namely Direct and Large Eddy Simulation.

Section 3.1.1 focuses on Direct Numerical Simulation (DNS). The section contains a discussion of the computational requirements for DNS of isotropic turbulence. The discussion shows that for high Reynolds number DNS is not possible due to high computational requirements.

Section 3.1.2 focuses on Large Eddy Simulation (LES), a simulation technique which is computationally less expensive but also less accurate than DNS. LES always needs a fluid-LES model, i.e., some assumption on the flow or the flow's statistics. All LES results presented in this work are based on the model proposed by Meneveau *et al.* (1996). This model is also presented in section 3.1.2.

For the sake of completeness it should be mentioned that there exists a third major technique for simulation of turbulent flow, which is Reynolds Averaged Navier–Stokes simulation (RANS). In RANS, one solves for averaged quantities only. Evidently, RANS is not very interesting for isotropic turbulence because here all averages are zero. Therefore, no RANS was conducted in the present work.

3.1.1 Direct Numerical Simulation (DNS)

Solving the Navier–Stokes equations numerically 'as is', is referred to as Direct Numerical Simulation (DNS). DNS with present numerical methods produces very reliable results such that DNS can be regarded as a substitute for an experiment. However, this holds only under the condition where DNS is possible. Actually for most cases DNS is not possible because the computational requirements for DNS often exceed computational ressources by far (cf. section 1.1).

The computational requirements for DNS are determined by the Reynolds number. For isotropic turbulence, the memory requirements can be estimated as follows. The size of the smallest scales is given by the Kolmogorov length scale η_K. This determines the cell size. However, in order to exactly represent the large scales, the computational domain would have to be infinitely large. The best approximation is a large box with periodic boundary conditions. This introduces artificial two point correlation of the fluid velocity at the scale of the box size. However, if the box is sufficiently large with respect to the integral length scale L_f, then this correlation is negligible. Pope (2000) recommends to use a computational box of 8 integral length scales, Reynolds (1990) recommends 2 integral length scales. Figure 2.1 indicates that, at least for the Reynolds number shown, the recommendation of 8 integral length scales is rather on the safe side.

Concluding, the ratio between the largest scales $8L_f$ and the smallest scales η_K determines the size of the computational box and the grid spacing, respectively. The model spectrum from section 2.1.3 predicts that $L_f/L_k \to 0.43$ for $Re_{L_k} \to \infty$. Thus, for high Reynolds number

$$\frac{L_f}{\eta_K} \approx 0.43 \frac{L_k}{\eta_K} = 0.43 Re_{L_k}^{3/4} \tag{3.1}$$

can be assumed. This result leads to the statement that the number of grid points (and thus memory) needed for DNS scales with $(Re_{L_k}^{3/4})^3 = Re_{L_k}^{9/4}$.

CPU time scales with the number of grid points and with the number of time steps required to obtain a statistically steady state. The number of time steps scales with $Re_{L_k}^{3/4}$ (cf. Pope, 2000). This means that CPU time scales with $Re_{L_k}^3$.

This shows that memory and run time requirements for the numerical computation of the solution to the Navier–Stokes equations (2.1) increase with Reynolds number. Actually, in most turbulent flows present in nature and industrial applications, the Reynolds number is so high that computational ressources are exceeded by far. For these configurations, the solution to equation (2.1) cannot be computed. A computationally less expensive but also less accurate alternative to DNS is Large Eddy Simulation (LES), elaborated in the following section.

3.1.2 Large Eddy Simulation (LES)

In the previous section it was shown that at high Reynolds number DNS is not possible because of the immense computational requirements associated with the wide range of scales contained in the flow. The fast decay of the energy spectrum function (cf. section 2.1.3) inspires to solve for large scales only, leading to Large Eddy simulation (LES). LES permits the use of a relatively coarse grid, i.e., to reduce computational requirements. On the other hand, LES is not as accurate as DNS. In the following, LES is explained in detail.

In LES, it is not the Navier–Stokes equations but rather filtered Navier–Stokes equations that are solved. This means that one does not ask for the fluid velocity but for the filtered fluid velocity. The filter is a spatial low pass filter. This allows the representation of the filtered velocity on a coarse grid.

The filtered Navier–Stokes equations are presented below. It will turn out that these are unclosed, i.e., not all terms can be expressed via filtered fluid velocity and filtered pressure only. Instead, an additional term, referred to as subgrid scale stress tensor τ, appears. This tensor must be modelled in terms of the filtered velocity. The tensor τ represents the effect of the unresolved scales on the resolved scales. In the present work, models for τ are referred to as *fluid-LES model*, cf. section 1.2. In this thesis, the fluid-LES model proposed by Meneveau et al. (1996) is implemented.

Each fluid-LES model implicitly defines a filter. This filter can be described a posteriori, i.e., by comparison of LES and DNS data. One compares the energy spectrum function from LES with the energy spectrum function from DNS and characterises the filter with respect to its damping properties of the energy spectrum function.

In the following, first the filtered Navier–Stokes equations are presented. Then, the fluid-LES model of Meneveau et al. (1996) is explained and finally transfer functions of the model and of standard filters are shown.

Filtered Navier–Stokes equation

Assuming that the exact fluid velocity $\mathbf{u}_f$ is known, one can compute the LES velocity $\langle \mathbf{u}_f \rangle$ by 'kicking out the small eddies'. Formulated more precisely, this means filtering $\mathbf{u}_f$ by a low-pass filter.

In LES, only linear filters are considered. Such a filter operator $\mathcal{G}$ can be defined by its kernel function $G(\mathbf{s})$,

$$\langle \mathbf{u} \rangle = \mathcal{G}\mathbf{u}(\mathbf{x}, t) = \iiint \mathbf{u}(\mathbf{x} - \mathbf{s}, t) G(\mathbf{s}) \, d\mathbf{s}. \tag{3.2}$$

Apparently for all filters defined via equation (3.2), filtering and differentiation commutes,

$$\frac{\partial \langle \cdot \rangle}{\partial t} = \left\langle \frac{\partial \cdot}{\partial t} \right\rangle, \quad \frac{\partial \langle \cdot \rangle}{\partial x_i} = \left\langle \frac{\partial \cdot}{\partial x_i} \right\rangle. \tag{3.3}$$

Therefore, if one applies the filter operator $\langle \cdot \rangle$ on the Navier–Stokes equations (2.1), then one ends up with the filtered Navier–Stokes equations (LES equations)

$$\frac{\partial \langle u_{f,i} \rangle}{\partial x_i} = 0 \tag{3.4a}$$

$$\frac{\partial \langle u_{f,i} \rangle}{\partial t} + \langle u_{f,j} \rangle \frac{\partial \langle u_{f,i} \rangle}{\partial x_j} + \frac{\partial \tau_{ij}}{\partial x_j} = -\frac{1}{\rho} \frac{\partial \langle p \rangle}{\partial x_i} + \nu \frac{\partial^2 \langle u_{f,i} \rangle}{\partial x_j^2}. \tag{3.4b}$$

Equations (3.4) are solved for $\langle \mathbf{u}_f \rangle$, the LES velocity. By definition, this is the fluid velocity which consists of large scales only and is therefore representable on a coarse grid. The tensor τ is defined by

$$\tau = \langle u_{f,i} u_{f,j} \rangle - \langle u_{f,i} \rangle \langle u_{f,j} \rangle. \tag{3.5}$$

3 Implemented simulation tools

τ is called *subgrid stress tensor*. This tensor represents the interaction between resolved large scales and unresolved small scales. τ depends on $\langle u_{f,i} u_{f,j} \rangle$. Therefore τ cannot be computed from the LES velocity directly but one needs an additional model equation. Here, this type of model is referred to as *fluid-LES model*. Each fluid-LES model implicitly defines a filter $\mathcal{G}$.

In the present work only one fluid-LES model was implemented, namely the Lagrangian Smagorinsky model proposed by Meneveau, Lund & Cabot (1996), presented in the next section.

The implemented fluid-LES model (Lagrangian Smagorinsky model)

In all Large Eddy Simulations presented in this work (except for simulations of other authors), τ was modelled according to the Lagrangian dynamic Smagorinsky model proposed by Meneveau *et al.* (1996). The model is based on

1. modelling τ using the eddy viscosity hypothesis (see Boussinesq, 1877; Smagorinsky, 1963)

2. estimation of the modelling error by the Germano identity (see Germano *et al.*, 1991)

3. minimisation of the estimated model error along trajectories of fluid particles.

These three points are briefly described one by one in the following.

The eddy viscosity hypothesis stems from the observation that large (resolved) eddies break up and generate small (unresolved) eddies. The unresolved eddies cannot be represented on the LES grid and therefore the breakup of large eddies can be modelled as dissipation. This means that τ can be modelled as dissipative term, i.e.,

$$\tau_{ij} = -2\nu_t \langle S_{ij} \rangle, \qquad S_{ij} = \frac{1}{2}\left(\frac{\partial u_{f,i}}{\partial x_j} + \frac{\partial u_{f,j}}{\partial x_i}\right) \qquad (3.6)$$

with the *eddy viscosity* ν_t. Such models are called *eddy viscosity models*. The first eddy viscosity model for LES was published by Smagorinsky (1963). He proposed

$$\nu_t = (C_S \Delta)^2 S \qquad \text{with } S = \sqrt{2 \langle S_{ij} \rangle \langle S_{ij} \rangle}. \qquad (3.7)$$

Δ denotes the filter associated with the LES grid. For a cuboidal cell with dimensions $\Delta x \times \Delta y \times \Delta z$ one can take for example

$$\Delta = (\Delta x \, \Delta y \, \Delta z)^{1/3}. \qquad (3.8)$$

C_S is a model constant, referred to as 'Smagorinsky constant'. Lilly (1967) considered the application of the Smagorinsky model on isotropic turbulence. Based on the spectrum of isotropic turbulence, he proposes to set

$$C_S = 0.17. \qquad (3.9)$$

However, it is clear that arbitrary filtered DNS fields will not show a constant value for C_S. The consequent refinement is to choose C_S in dependence of the resolved scales. Such models are called dynamic Smagorinsky models. Meneveau, Lund & Cabot (1996) proposed a specific dynamic Smagorinsky model, making use of the Germano identity.

Germano et al. (1991) had the following idea: The LES equations correspond to filtered Navier–Stokes equations. If one filters the LES equations again, then one obtains double-filtered Navier–Stokes equations

$$\frac{\partial \langle\langle u_{f,i}\rangle_1\rangle_2}{\partial x_i} = 0 \tag{3.10a}$$

$$\frac{\partial \langle\langle u_{f,i}\rangle_1\rangle_2}{\partial t} + \langle\langle u_{f,j}\rangle_1\rangle_2 \frac{\partial \langle\langle u_{f,i}\rangle_1\rangle_2}{\partial x_j} + \frac{\partial T_{ij}}{\partial x_j} = -\frac{1}{\rho}\frac{\partial \langle\langle p\rangle_1\rangle_2}{\partial x_i} + \nu \frac{\partial^2 \langle\langle u_{f,i}\rangle_1\rangle_2}{\partial x_j^2}. \tag{3.10b}$$

Here, $\langle \cdot \rangle_1$ and $\langle \cdot \rangle_2$ are two filters and $\mathbf{T}$ is the subgrid stress tensor obtained from double-filtering, $T_{ij} = \langle\langle u_{f,i} u_{f,j}\rangle_1\rangle_2 - \langle\langle u_{f,i}\rangle_1\rangle_2 \langle\langle u_{f,j}\rangle_1\rangle_2$. If one applies $\langle \cdot \rangle_2$ on τ, then one obtaines the Germano identity (see Germano et al., 1991; Germano, 1992):

$$T_{ij} - \langle \tau_{ij}\rangle_2 = \langle\langle u_{f,i}\rangle_1 \langle u_{f,j}\rangle_1\rangle_2 - \langle\langle u_{f,i}\rangle_1\rangle_2 \langle\langle u_{f,j}\rangle_1\rangle_2 \qquad \text{(Germano identity)} \tag{3.11}$$

This identity can be used to estimate the error of a LES model. In LES, $\langle \mathbf{u}_f\rangle_1$ is computed. $\langle \cdot \rangle_1$ is implicitly defined by the fluid-LES model. The second filter $\langle \cdot \rangle_2$ is defined explicitly. This allows to compute the right hand side of equation (3.11). The terms on the left hand side can be computed by implementing the LES model. If the filter $\langle \cdot \rangle_2$ is suited to the model, then the difference between (explicitly computed) right hand side and (modelled) left hand side gives an idea of the model error.

The Germano identity can also be used to set model parameters. For example, one can use an eddy viscosity model for τ and $\mathbf{T}$, and use the Germano identity in order to set the Smagorinsky constant C_S. However, this gives no good value for C_S because this argumentation neglected the fact that in general one can only expect a LES model to work in a statistical sense. Therefore, averaging is necessary. Germano et al. (1991) propose to average in homogeneous directions. If the flow is not homogeneous, then this is evidently not possible.

One workaround to that was proposed by Meneveau et al. (1996). They propose to average along trajectories of fluid particles. The authors furthermore respect that for inhomogeneous flows it does not make sense to average along the whole particle path because the statistics of the flow seen by the particle are not constant. Therefore Meneveau et al. (1996) additionally introduce a weighting function such that the recently seen flow has a stronger influence on C_S than the flow which the particle saw long time ago. More precisely, Meneveau et al. (1996) propose an exponential weighting function, i.e., at some instant t_0, the error estimation from $t_0 - t$ is weighted by $e^{(t-t_0)/t^{LES}}$. The normalisation factor t^{LES} is set in dependence of the estimated error.

Putting it all together, Meneveau et al. (1996) set C_S as follows:

3 Implemented simulation tools

- Denote by $\overline{\cdot}^M$ averaging along particle's trajectories,

$$\overline{f(t,\mathbf{x})}^M = \frac{1}{t_{LES}} \int_{-\infty}^{t} f(t',\mathbf{x}_p(t')) \, e^{\frac{t'-t}{t_{LES}}} \, dt' \tag{3.12}$$

where $\mathbf{x}_p(t)$ is the path of a particle which is transported with the LES velocity $\langle \mathbf{u}_f \rangle$. t_{LES} is defined below.

- Define the test filter $\langle \cdot \rangle_2$. Meneveau et al. (1996) propose a sharp spectral filter at twice the grid scale. In the present work, a box filter at twice the grid scale was implemented for reasons of computational efficiency. Both filters are defined below.

- Compute in each time step the right hand side of Germano's identity

$$L_{ij} = \langle\langle u_{f,i}\rangle \langle u_{f,j}\rangle\rangle_2 - \langle\langle u_{f,i}\rangle\rangle_2 \langle\langle u_{f,j}\rangle\rangle_2 \tag{3.13}$$

- Replace the left hand side of Germano's identity according to equations (3.6) and (3.7),

$$T_{ij} - \langle\tau_{ij}\rangle_2 = C_S^2 M_{ij}, \tag{3.14a}$$
$$M_{ij} = 2\Delta^2 \langle S \langle S_{ij}\rangle\rangle_2 - 2\Delta_2^2 S^{(2)} \langle\langle S_{ij}\rangle\rangle_2, \tag{3.14b}$$
$$S^{(2)} = \sqrt{2\langle\langle S_{ij}\rangle\rangle_2 \langle\langle S_{ij}\rangle\rangle_2} \tag{3.14c}$$

- Compute

$$t^{LES} = 1.5\Delta \left(\overline{L_{ij}M_{ij}}^M \, \overline{M_{ij}M_{ij}}^M \right)^{-1/8}. \tag{3.15}$$

- Choose C_S such that the averaged square error, computed from Germano's identity, is minimised:

$$\overline{(L_{ij} - C_S^2 M_{ij})^2}^M \to \min. \tag{3.16a}$$

This is equivalent to

$$\frac{d}{dC_S} \overline{(L_{ij} - C_S^2 M_{ij})^2}^M = 0, \quad \text{i.e.,} \quad C_S^2 = \frac{\overline{L_{ij}M_{ij}}^M}{\overline{M_{ij}M_{ij}}^M}. \tag{3.16b}$$

This formulation is complete but it is computationally expensive because equation (3.12) requires infinite backward integration. Meneveau et al. (1996) circumvent this by formulation of differential equations for $\overline{L_{ij}M_{ij}}^M$ and $\overline{M_{ij}M_{ij}}^M$. Furthermore, equation (3.15) shows that t_{LES} depends on averaged values. The time scale for averaging is t_{LES} again. Thus, equation (3.15) is an implicit equation. Meneveau et al. (1996) propose to compute $\overline{L_{ij}M_{ij}}^M$ and $\overline{M_{ij}M_{ij}}^M$ by using a value for t_{LES} which was computed in the previous time step and then update t_{LES} by equation (3.15).

In comparison to the standard Smagorinsky model, the dynamic approach has turned out to be more accurate (cf. e.g. Vreman et al., 1997; Meneveau & Lund, 1997) but of course computationally more expensive. Kuerten et al. (1999) proposed an approach which is less expensive than the dynamic model but seems to perform comparably well.

However, in the present study only the Lagrangian Smagorinsky model was implemented because with this model experience from several previous studies exists (cf. e.g. Gobert et al., 2007; Gobert & Manhart, 2007; Link et al., 2008; Gobert & Manhart, 2009; Gobert et al., 2010). Actually one can expect that for particle-laden flow this model is very well suited because the model is constructed such that the estimated error is minimised along a particle's path.

Characterisation of a fluid-LES model via filter transfer functions

The quality of a fluid-LES model can be analysed via its filter kernel G and the corresponding Fourier transform $\mathcal{FT}(G)$. The latter function is also called *filter transfer function*. A somewhat high quality fluid-LES model would correspond to a filter which does not touch the large scales and removes all scales smaller than a certain limit. More precisely, the filter transfer function $\mathcal{FT}\left(G^{Sharp}\right)$ of such a model would be a Heaviside function,

$$\left(\mathcal{FT}\left(G^{Sharp}\right)\right)(\mathbf{k}) = \begin{cases} 1 & \text{if } \|\mathbf{k}\| < \kappa_c \\ 0 & \text{if } \|\mathbf{k}\| \geq \kappa_c \end{cases}. \tag{3.17}$$

κ_c is called cutoff wavenumber and $\mathcal{G}^{Sharp}$ is called sharp spectral filter. In a simplified point of view, all eddies larger than $2\pi/\kappa_c$ are not modified by $\mathcal{G}^{Sharp}$ and all eddies smaller than $2\pi/\kappa_c$ are removed by $\mathcal{G}^{Sharp}$.

Other commonly used filters are the box filter or the Gaussian filter, defined by

$$G^{box}(\mathbf{s}) = \begin{cases} \frac{\kappa_c}{\pi} & \text{if } \|\mathbf{s}\| \leq \frac{\pi}{\kappa_c} \\ 0 & \text{otherwise} \end{cases}, \quad G^{Gauss}(\mathbf{s}) = \frac{\sqrt{6}\kappa_c}{\pi^{3/2}} \exp\left(-\frac{6\|\mathbf{s}\|^2\kappa_c^2}{\pi^2}\right). \tag{3.18}$$

For details on the choice of the coefficients, the reader is referred to Pope (2000). The transfer function of these filters are depicted in figure 3.1. The x-axis shows $\kappa/\kappa_c = \|\mathbf{k}\|/\kappa_c$.

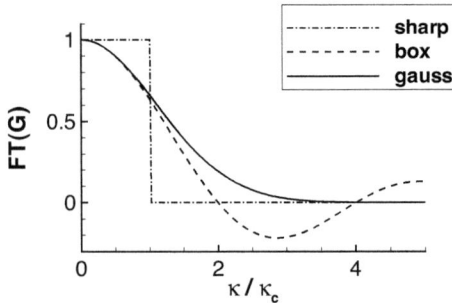

Figure 3.1: Filter transfer functions of sharp spectral, box and Gaussian filter.

3 Implemented simulation tools

From this point of view one might think that a somewhat 'perfect' filter is the sharp spectral filter because here, all resolvable wavenumbers are resolved exactly. But the sharp spectral filter also has disadvantages. Because of $G^{box} = \kappa_c/\pi \left(\mathcal{FT} \left(G^{Sharp} \right) \right)$, the function G^{Sharp} is identical to $\pi/\kappa_c \mathcal{FT}^{-1} \left(G^{box} \right)$,

$$G^{Sharp} = \frac{\pi}{\kappa_c} \mathcal{FT}^{-1} \left(G^{box} \right) = \frac{\pi}{\kappa_c} \mathcal{FT} \left(G^{box} \right). \tag{3.19}$$

The latter equality holds due to the symmetry of G^{box}.

$\mathcal{FT} \left(G^{box} \right)$ is depicted in figure 3.1. This shows that G^{Sharp} has negative loops. This means that the sharp spectral filter may invert the velocity, i.e., filtering of a positive velocity component may lead to a negative component. Evidently this property is undesired. Thus, it is not clear which filter is the best.

It remains to establish a link between a specific model for the subgrid stress tensor τ and the corresponding filter $\mathcal{G}$. This link can be obtained via Fourier transformation:

$$\left(\mathcal{FT}(G) \right)(\mathbf{k}) = \frac{\left(\mathcal{FT} \left(\langle \mathbf{u}_f \rangle \right) \right)(\mathbf{k})}{\left(\mathcal{FT}(\mathbf{u}_f) \right)(\mathbf{k})}. \tag{3.20}$$

For isotropic filters the correspondent filter transfer function depends only on the norm of $\mathbf{k}$, i.e., $\left(\mathcal{FT}(G) \right)(\mathbf{k}) = \left(\mathcal{FT}(G) \right)(\|\mathbf{k}\|)$. Then one can compute the filter transfer function from the energy sprectrum functions,

$$\left| \left(\mathcal{FT}(G) \right)(\|\mathbf{k}\|) \right|^2 = \frac{E^{\langle \cdot \rangle}(\|\mathbf{k}\|)}{E(\|\mathbf{k}\|)}. \tag{3.21}$$

Here, $E^{\langle \cdot \rangle}$ and E denote the energy spectrum functions of filtered and unfiltered velocity, respectively. For a given fluid-LES model, one computes $E^{\langle \cdot \rangle}$ by LES and E by DNS.

An example for a filter transfer function is depicted in figure 3.2. The underlying fluid-LES model is the Lagrangian dynamic Smagorinsky model. The transfer function was computed from equation (3.21). $E^{\langle \cdot \rangle}$ and E were computed from instantaneous LES and DNS data. The underlying flow is homogeneous isotropic turbulence at $Re_\lambda = 99$. Evidently, the transfer functions of box filter and Gaussian filter are different but comparable.

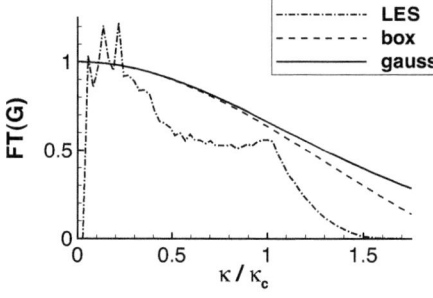

Figure 3.2: Filter transfer function of the Lagrangian Smagorinsky model from section 3.1.2 together with the transfer function of the box filter and the Gaussian filter. For the Lagrangian Smagorinsky model the transfer function was computed via LES and DNS of isotropic turbulence at $Re_\lambda = 99$.

Apparently the transfer function of the Lagrangian Smagorinsky model shows a kink at $\kappa = \kappa_c$. This can be explained as follows. The simulation was performed on a cube discretised by an equidistant Cartesian grid. Therefore the wavenumbers which are resolved in LES also form an equidistant Cartesian grid within a cube $[-\kappa_c, \kappa_c]^3$ in wavenumber space. For $\kappa_c < \kappa \leq \sqrt{3}\kappa_c$, a sphere with radius κ intersects the cube but does not fit within the cube. $E^{\langle \cdot \rangle}(\kappa)$ is computed by integration over the surface of such a sphere and therefore $E^{\langle \cdot \rangle}(\kappa)$ decays between κ_c and $\sqrt{3}\kappa_c$ simply because the corresponding wavenumbers are only partially resolved.

3.2 Numerical methods for the carrier fluid

The previous section focused on simulation techniques. DNS and LES was explained. Governing equations were presented but no methods for numerical discretisation were discussed. Furthermore, it was not discussed how to simulate isotropic turbulence.

Both topics are addressed in the following. First, the implemented methods for discretisation of the Navier–Stokes and filtered Navier–Stokes equations are briefly listed. Then, the implemented method for generation of isotropic turbulence is presented.

3.2.1 Basic numerical scheme

In DNS the Navier–Stokes equations are solved, in LES one solves the filtered Navier–Stokes equations. In the present work the same numerical scheme was implemented for both equations.

This scheme is implemented in the flow solver 'MGLET'. MGLET is an in-house code of Fachgebiet Hydromechanik at TU München. It was developed and successfully used by Manhart (1995, 2004), Manhart et al. (2001); Manhart & Friedrich (2002) and Peller et al. (2006). The referenced studies contain details on the flow solver. In the following only the most relevant characteristics are listed.

Discretisation in space is achieved by a second-order Finite Volume scheme. In particular, this means that one solves for volume or surface averaged quantities. This peculiarity will be addressed again in section 3.3.1, the interpolation of the fluid velocity on the particle position.

The variables are arranged on a staggered grid. As mentioned previously, the only testcase for the present work is isotropic turbulence. Therefore only equidistant Cartesian grids were used.

Time advancement was implemented by a third-order Runge–Kutta scheme proposed by Williamson (1980). Incompressibility was enforced by solving the Poisson equation for the pressure using an iterative solver proposed by Stone (1968).

Altogether the flow solver consists of standard methods, for example described by Ferziger & Peric (1999) and Kundu & Cohen (2004).

3.2.2 Forcing of isotropic turbulence

The previous section described how the Navier–Stokes and filtered Navier–Stokes equations were discretised. The present section explains how the flow was driven in order to obtain isotropic turbulence.

Probably one could write a book on 'how to generate isotropic turbulence'. This is an old issue, discussed in many studies. In this work only a brief summary is given. The aim is to give the reader an idea on what configuration one could think of when referring to isotropic turbulence.

In experiments, isotropic turbulence is often produced by passing a flow through a grid. At a sufficiently large distance behind the grid, turbulent fluctuations are quasi homogeneous and isotropic.

For particle-laden flow, an experimental setup of isotropic turbulence is more difficult because for inert particles gravity always leads to anisotropies for the particles. Therefore Hwang & Eaton (2006) conducted an experiment in parabolic flight.

Admittedly, the above experiment is rather exceptional and the experimentators are to be envied. For numerical simulation, the generation of isotropic tubulence is much less spectacular. Here, one usually solves the Navier–Stokes equations in a cube with periodic boundary conditions. The necessary dimensions of the cube are discussed in section 3.1.1.

One distinguishes between two types of isotropic turbulence, namely decaying and forced isotropic turbulence. Decaying isotropic turbulence corresponds to grid generated turbulence. If one travels with the flow in grid generated turbulence, then one observes that the turbulent fluctuations decay due to dissipation. In a numerical simulation, such a process can be emulated by specification of the flow at some initial instant and solving the Navier–Stokes equations (2.1) until the flow is at rest. The initial flow is typically either taken from experiments or set in accordance with a model spectrum (cf. chapter 2.1.3).

The initial flow field significantly determines the whole process. The statistics of the flow field are unsteady and the Reynolds number decays. This makes temporal statistical sampling different to spatial statistical sampling.

In order to increase the number of samples, one aims at a statistically steady field. This is achieved by simulation of so-called forced isotropic turbulence. Forced isotropic turbulence stands for isotropic turbulence with constant application of energy, called forcing. Following Kolmogorov's hypotheses, it makes sense to apply energy only at low wavenumbers. Then, statistics of high wavenumbers should correspond to statistics in an arbitrary flow.

Forcing is established by an additional term in the Navier–Stokes equations. This term can be either a stochastic process or it can be deterministic. Correspondingly, one distinguishes between deterministic and stochastic forcing schemes. Deterministic forcing schemes typically tend to produce anisotropic flow fields whereas stochastic schemes usually lead to long-term oscillations (see e.g. Overholt & Pope, 1998; Taylor et al., 2003). This means that with stochastic forcing schemes one needs to run simulations for a longer period in order to obtain reliable statistics. The other way round, this means that with the available computational resources a stochastic forcing scheme leads to less reliable statistics or restricts the simulations to lower Reynolds number. In order to circumvent this, in the present work a deterministic forcing scheme was chosen.

The implemented scheme is a slightly modified version of the forcing scheme proposed by Sullivan et al. (1994). As for most forcing schemes, Sullivan et al. (1994) start with a spectral description of the flow. They choose a large scale wavenumber κ_L and specify the energy budget contained in the large scales, $k_L = \frac{1}{2} \int_{\|\kappa\| < \kappa_L} \|\mathcal{FT}(\mathbf{u}_f)(\kappa)\|^2 \, d\kappa$. Within one time step, this energy budget will decrease due to energy transfer to smaller scales and due to dissipation. Therefore the authors propose to enhance after each time step the components $\mathcal{FT}(\mathbf{u}_f)(\kappa)$, $\|\kappa\| < \kappa_L$ by a wavenumber-independent multiplicative factor α such that $\frac{1}{2} \int_{\|\kappa\| < \kappa_L} \|\alpha \mathcal{FT}(\mathbf{u}_f)(\kappa)\|^2 \, d\kappa$ equals the specified energy k_L.

Thus, the scheme consists of two parameters, namely k_L and κ_L. This scheme was implemented by various authors (see e.g. Sullivan et al., 1994; Bogucki et al., 1997; Vaillancourt et al., 2002). Their simulations and also own simulations showed that the scheme does not lead to an increase of energy in the low wavenumber range, i.e., the spectrum's peak coincides with the lowest resolved wavenumber. This is in contrast to the spectrum one obtains using e.g. the forcing schemes by Eswaran & Pope (1988) or Overholt & Pope (1998). In order to facilitate comparison with studies using one of these forcing schemes, in the present work only the modes in a given range $[\kappa_0, \kappa_L]$ were forced. Thus, in contrast to the scheme proposed by Sullivan et al. (1994), the modes at the lowest resolved wave numbers are not forced. This leads to a decay in the spectrum at the smallest wavenumbers. The factor α was found to range typically between 1.001 and 1.01.

3.3 Numerical methods for particle-laden flow

The previous section contained numerical methods for DNS and LES of single phase flow. The present section contains the counterparts for computation of the particle dynamics.

In addition to the Navier–Stokes equations for the carrier flow, for each particle the particle transport equation (2.32) must be solved. To this end, the fluid velocity must be interpolated at the particle's position.

Interpolation is discussed in section 3.3.1. In section 3.3.2, a numerical method for solving equation (2.32) is presented. This method was used for all simulations presented in this work.

The methods presented are based on studies of Yeung & Pope (1988), Balachandar & Maxey (1989), Meyer & Jenny (2004), Rosenbrock (1963), Kaps & Rentrop (1979) and Gottwald & Wanner (1981).

3.3.1 Interpolation of the fluid velocity on the particle's position

In order to compute the particle velocity (equation (2.32)), one needs to compute the fluid velocity seen by the particle at every time step. With a second-order Finite Volume method, cell volume or cell surface averaged values for the fluid velocity are computed. Averaged values can be regarded as filtered values. The corresponding filter is a box filter (cf. section 3.1.2). On the other hand, the fluid velocity seen by a particle is a point value. Therefore,

3 Implemented simulation tools

strictly speaking, in order to compute the fluid velocity seen by a particle at a random position one would need to 'invert' the averaging operator (i.e., 'defilter' the field) first and then interpolate the defiltered field on the particle's position.

Of course, the averaging operator is not invertible and therefore the problem cannot be tackled that way. In LES, defiltering is addressed by particle-LES models (cf. chapters 6 and 7). In DNS, this is not a big issue because if the computational cells are smaller than the Kolmogorov length then surface averaged values do not differ significantly from point values. In this case, one can treat the results of a Finite Volume simulation as point values and obtain the fluid velocity seen by the particle by interpolation.

Concerning interpolation of fluid velocity on particle position, there are three major studies to mention, namely the studies of Yeung & Pope (1988), Balachandar & Maxey (1989) and Meyer & Jenny (2004). These studies are typically referenced in order to select a scheme for interpolation (see e.g. Toschi & Bodenschatz, 2009).

Yeung & Pope (1988) and Balachandar & Maxey (1989) analysed different interpolation schemes for inertia free particles. They conducted an analysis in homogeneous isotropic turbulence and compared polynomial interpolation of various order, cubic splines and trigonometric interpolation against each other. Trigonometric interpolation was assumed to give exact values because the flow is decomposable on a Fourier basis. With this reference solution they computed the error for the other schemes from the differences in particle position.

They found that the error from second-order interpolation is significantly higher than the error from third-order interpolation. Fourth-order shows smaller errors than third-order but the difference between third and fourth-order is not substantial. The error obtained with splines turned out to be comparable to third-order polynomial interpolation, but polynomial interpolation is computationally less expensive than splines due to locality. Therefore the authors recommend the use of third or fourth-order interpolation.

Meyer & Jenny (2004) addressed another aspect, namely artificial clustering due to interpolation errors. This occurs if the interpolated fluid velocity is not divergence free. They showed that standard non-conservative second-order interpolation schemes lead to substantial artificial clustering.

Unless noted otherwise, the data presented in this work is based on a standard fourth-order interpolation scheme, meeting the requirements stated by Yeung & Pope (1988) and Balachandar & Maxey (1989). Additionally, simulations were conducted using a second-order conservative scheme. This scheme is explained in detail in Gobert *et al.* (2006). It meets the requirements of Meyer & Jenny (2004). In DNS, both schemes produced very similar results, the effect of interpolation is negligible due to the fine grid. In LES, an effect was observable but it is so small that it does not affect the overall conclusions. Special attention was paid to the effect of interpolation on preferential concentration, sections 5.5.4, 6.4.4 and 7.4.3. Also here, results from fourth-order non-conservative and second-order conservative interpolation do not differ substantially. Therefore the results from chapters 4 to 6 are based on fourth-order interpolation unless noted otherwise. An analysis of the effect of interpolation can be found in chapter 7, section 7.1.1.

3.3.2 Computation of particle velocity

After selecting an interpolation method for computing $\mathbf{u}_{f@p}$, one can tackle the particle transport equation (2.32). In the following it is described how this was conducted in the present work.

In the implemented approach, in each time step of the flow solver the velocity of each particle is advanced accordingly. In other words, all particles and flow are synchronised after each time step of the flow solver.

For small Stokes numbers, Stokes drag is a stiff term. In general, stiff differential equations are computationally expensive because explicit schemes require very small time steps and implicit schemes require the solution of a system of non-linear equations. One exception are linear stiff equations. Here, implicit schemes reduce to the solution of a system of linear equations while allowing for large time steps.

Rosenbrock–Wanner schemes (see Hairer & Wanner, 1990; Deuflhard & Bornemann, 2008) are schemes for stiff equations which combine an implicit and an explicit approach. The idea is to decompose the differential equation in a linear and a non-linear term and to discretise the linear term implicitly and the non-linear term explicitly. More precisely, in the context of the present work, the particle transport equation (2.32) is first rewritten in the form

$$\frac{du_{p,i}}{dt} = f_i(t, \mathbf{u}_p), \qquad f_i(t, \mathbf{u}_p) = \frac{c_D Re_p}{24 \tau_p}(u_{f@p,i} - u_{p,i}) \qquad (3.22a)$$

$$\frac{du_{p,i}}{dt} = \frac{\partial f_i}{\partial t} t + \frac{\partial f_i}{\partial u_{p,j}} u_{p,j} + \left(f_i(t, \mathbf{u}_p) - \frac{\partial f_i}{\partial t} t - \frac{\partial f_i}{\partial u_{p,j}} u_{p,j} \right). \qquad (3.22b)$$

Then, the first terms of equation (3.22b), $\frac{\partial f_i}{\partial t}t + \frac{\partial f_i}{\partial u_{p,j}}u_{p,j}$, are discretised by an implicit Runge-Kutta scheme and the remaining terms by an explicit Runge-Kutta scheme. It should be mentioned that $\frac{\partial f_i}{\partial t}$ is not zero because $\mathbf{u}_{f@p}$ depends on t.

In order to obatin a first order Rosenbrock–Wanner scheme one writes

$$u_{p,i}(t + \Delta t) \doteq u_{p,i}(t) + \Delta t \left(\frac{\partial f_i}{\partial t} \Delta t + \frac{\partial f_i}{\partial u_{p,j}} u_{p,j}(t + \Delta t) \right)$$
$$+ \Delta t \left(f_i(t, \mathbf{u}_p(t)) - \frac{\partial f_i}{\partial u_{p,j}} u_{p,j}(t) \right) \qquad (3.23)$$

and then sort this equation with respect to knowns and unknowns,

$$\left(\delta_{ij} - \frac{\partial f_i}{\partial u_{p,j}} \right) u_{p,j}(t + \Delta t) \doteq u_{p,i}(t) + \frac{\partial f_i}{\partial t} \Delta t^2$$
$$+ \Delta t \left(f_i(t, \mathbf{u}_p(t)) - \frac{\partial f_i}{\partial u_{p,j}} u_{p,j}(t) \right). \qquad (3.24)$$

Δt denotes the time step size of the flow solver, δ is the Kronecker delta function and $\doteq$ stands for first-order approximation. The derivatives of $\mathbf{f}$ are evaluated at $(t, \mathbf{u}_p(t))$. It should be noted that c_D and Re_p depend on $\mathbf{u}_p$ and therefore $\frac{\partial f_i}{\partial u_{p,j}}$ is not a constant.

Equation (3.24) shows that with this discretisation $\mathbf{u}_p(t + \Delta t)$ can be approximated by

3 Implemented simulation tools

solving the system of linear equations $\left(\delta_{ij} - \frac{\partial f_i}{\partial u_{p,j}}\right)$. In the present case this is inexpensive because the system is only three-dimensional. Furthermore, the linear approximation is very well suited for the particle transport equation (2.32) because in the low Reynolds number limit $(Re_p \to 0)$, $c_D = 24/Re_p$, Stokes drag is linear and therefore the linear approximation is exact. Also for high Reynolds number, the approximation is good. For example Schiller and Naumann's model for c_D gives only a maximal exponent of 1.687 for $\mathbf{u}_p$, i.e., Stokes drag is not even quadratic.

Equation (3.24) is a first order discretisation. In the present study a fourth-order Runge-Kutta scheme was used instead. In this case, the right hand side $\mathbf{f}$ must be evaluated at various instants between t and $t+\Delta t$. However, the flow solver provides $\mathbf{u}_{f@p}$ only at intervals of Δt. Therefore $\mathbf{u}_{f@p}(t)$ was interpolated linearly in time.

One might wonder about the effect of this interpolation on the overall accuracy. In the present work, no effect was observable. The time step size of the flow solver was so small that the interpolation error from this temporal interpolation is negligible.

This interpolation requires that $\mathbf{u}_{f@p}$ is known at the new particle's position, $\mathbf{x}_p(t + \Delta t)$. This again means that first the fluid must be advanced to $t + \Delta t$, then the particle must be transported to its new position $\mathbf{x}_p(t + \Delta t)$, then $\mathbf{u}_{f@p}(t + \Delta t)$ must be computed by spatial interpolation and finally $\mathbf{u}_p(t+\Delta t)$ can be computed as described. For advancing the particle to its new position an explicit Euler method was used,

$$\mathbf{x}_p(t + \Delta t) = \mathbf{x}_p(t) + \Delta t\,\mathbf{u}_p(t). \tag{3.25}$$

Rosenbrock–Wanner schemes furthermore require the computation of $\frac{\partial f_i}{\partial t}$ and $\frac{\partial f_i}{\partial u_{p,j}}$. In the present case, this was done analytically, respecting that c_D and Re_p depend on $\mathbf{u}_p$.

The implemented scheme is adaptive with a third-order error estimator, i.e., for the particles, the time step Δt was subdivided in smaller time steps such that the difference between the third-order solution and the fourth-order solution is smaller then an error bound. The error bound was set to $10^{-2}u_{p,i} + 10^{-4}$ for each component of $u_{p,i}$. This bound is particularly important if the Stokes number is small but greater than zero. Section 5.4 contains a comparison of simulated data against experimental data at $St = 0.1$. It shows that even at this Stokes number the error bound gives satisfactory results. In the present thesis no particles with Stokes numbers greater than zero and smaller than 0.1 were analysed. Therefore the error bound can be assumed to be sufficiently small for all Stokes numbers.

Although not conducted in the present work, Rosenbrock–Wanner schemes are also well suited if the particle transport equation includes the fluid acceleration force as well. Actually, linear Stokes drag and fluid acceleration force lead to the test equation proposed by Prothero & Robinson (1974). Bartel & Günther (2002) showed that Rosenbrock–Wanner schemes are well suited for this class of problems. If even more terms are included in the particle transport equation, then the performance of Rosenbrock–Wanner schemes can be analysed by extending the Prothero-Robinson test equation, in analogy to the analysis of Simeon (1998, 2001).

3.4 Summary of implemented numerical methods

Sections 3.1 to 3.3 of this chapter contained descriptions of the numerical tools implemented in this work. These are summarised in the present section.

Computation of the carrier flow (MGLET):

- Governing equations: Navier–Stokes (DNS) and filtered Navier–Stokes with fluid-LES model (LES)
- Fluid-LES model: Dynamic Lagrangian Smagorinsky model proposed by Meneveau et al. (1996)
- Discretisation of Navier–Stokes equations (DNS) and filtered Navier–Stokes equations (LES): Second-order Finite Volume scheme
- Grid: Staggered equidistant Cartesian grid
- Advancement in time: Third-order Runge–Kutta scheme proposed by Williamson (1980)
- Enforcement of Conservativity: Poisson equation for the pressure using an iterative solver proposed by Stone (1968)

Computation of the particles (additionally implemented in MGLET for the present work):

- Discretisation: Point-particles, Euler–Lagrange approach
- Coupling: One-way coupling
- Governing equation: Simplified Maxey–Riley equation, considering Stokes drag only
- Empirical corrections: Correction of the drag coefficient for high particle Reynolds numbers following the recommendations of Clift et al. (1978)
- Interpolation of fluid velocity on particle position: Fourth-order Lagrangian interpolation unless noted otherwise
- Time advancement of particle velocity: Adaptive fourth-order Rosenbrock–Wanner scheme with third-order error estimator
- Time advancement of particle position: Explicit Euler scheme

4 A methodology for assessment of particle-LES models

Never trust a statistic you didn't fake yourself.
Winston Churchill

Chapter 2 covered fundamental issues, chapter 3 focused on numerical tools. In chapters 5 to 7 these tools are used in order to analyse SGS effects and to assess particle-LES models. To this end, SGS effects must be separated from large scale effects and data must be evaluated statistically. The present chapter explains how this is conducted in chapters 5 to 7. Thus, the present chapter explains the methodology which is implemented in the remaining chapters.

This and the following chapters concern the main topic of this work, LES of particle-laden flow. Focus is on the effect of small scales on particles and its reconstruction by a particle-LES model.

For some applications this effect is negligible, for others it is crucial. Therefore, before implementing a particle-LES model, one should first clarify if a model is needed and, if yes, what properties the model should have. Therefore, the first question must be 'which quantities are of interest for the given application?'.

The present work addresses particle-laden flow in general. The quantities of interest are defined by a statistical approach. More precisely, the quantities of interest are first and second moments of particle position, particle velocity and fluid velocity seen by the particles. In addition, preferential concentration is taken into consideration. The present chapter focuses on a methodology to compute these quantities.

For the computation of the statistical moments, an averaging operator must be defined. This is conducted in section 4.1. That section also contains simplifications for the averaging operator for forced isotropic turbulence. Section 4.2 lists the quantities which will serve in chapters 5 to 7 for the analysis of small scale effects and for model assessment. In addition, transport equations for these moments are presented. Section 4.3 focuses on techniques to compute the quantities under consideration by numerical simulation with regard to small scale effects. In particular, different techniques for different purposes are explained. These techniques will be implemented in the following chapters.

4.1 Definition of an averaging operator for particles

In this section, an averaging operator for the particles is defined. This operator respects the Lagrangian nature of the particle dynamics. Therefore it is not equivalent to an average operator for the carrier fluid flow.

In forced isotropic turbulence, one can compute statistics of the carrier fluid flow by averaging over space and time (cf. section 2.1.1). Correspondingly, one can compute statistics of the suspended phase by averaging over particles and time. This holds for forced isotropic turbulence. In section 6.3 and 7.3, results from analytical computations will be presented which are not restricted to isotropic turbulence. On the contrary, that analysis also concerns general turbulent flow. For general flows it is not clear how to average quantities related to the particles. For example one option would be to record the particle-laden flow at some instant, divide the domain in small cubes and to average in each cube over all particles which reside in that cube. In this case, one identifies two particles with each other if they reside within the same cube at the recorded instant. Such an approach corresponds to the question 'What are the statistical properties of particles which arrive at a specific position at a specific instant?'. This is a Eulerian approach. Another option would be to identify two particles with each other if they were released within the same cube, independent of their position at the recorded instant. This approach corresponds to the question 'What will on average happen to a particle if I release it at a specific position at a specific instant?'. This is a Lagrangian approach.

In the following, the latter approach is used. Consider one single particle. Its position $\mathbf{x}_p(t)$ and velocity $\mathbf{u}_p(t)$ at some instant t depends on its initial position and velocity $\mathbf{x}_0 = \mathbf{x}_p(t_0)$ and $\mathbf{u}_{p,0} = \mathbf{u}_p(t_0)$. In addition, $\mathbf{x}_p(t)$ and $\mathbf{u}_p(t)$ depend on the flow.

Now, compute an ensemble average for the flow, i.e., average $\mathbf{x}_p(t)$ and $\mathbf{u}_p(t)$ over multiple realisations of the flow. Denote the corresponding quantities by $\overline{\mathbf{x}_p}(t)$ and $\overline{\mathbf{u}_p}(t)$.

More general, denote by $\overline{f}(t|\mathbf{x}_0, \mathbf{u}_{p,0})$ the ensemble averaged quantity f. f can be any operator which depends on $t, \mathbf{x}_0, \mathbf{u}_{p,0}$ and the (time dependent) functions $\mathbf{x}_p(t)$, $\mathbf{u}_p(t)$ and $\mathbf{u}_{f@p}(t)$. This means that f can be a primitive variable such as $f = \mathbf{x}_p$ or $f = \mathbf{u}_p$ but f can also be a non-primitive quantity such as the integral time scale of particle velocity.

Ensemble averaging over the flow means that one conducts a large number N of realisations of the flow, enumerated by $i = 1, 2, 3, \ldots, N$. For each realisation, one will obtain different results for the (time dependent) particle position, particle velocity and velocity seen by the particle. Enumerate these by $\mathbf{x}_p^i(t), \mathbf{u}_p^i(t)$ and $\mathbf{u}_{f@p}^i(t)$. Then, the ensemble average $\overline{f}(t|\mathbf{x}_0, \mathbf{u}_{p,0})$ can be expressed by

$$\overline{f}(t|\mathbf{x}_0, \mathbf{u}_{p,0}) = \frac{1}{N} \sum_{i=1}^{N} f\left(\mathbf{u}_{f@p}^i, \mathbf{u}_p^i, \mathbf{x}_p^i; t | \mathbf{x}_0, \mathbf{u}_{p,0}\right). \tag{4.1}$$

It should be mentioned that in general this averaging operator does not include averaging over time, space or particles. It stands only for ensemble averaging over the flow. This is important for the analytical computations in section 6.3 and 7.3.

In a statistically steady flow, ensemble averaging is equivalent to averaging in time. Nevertheless, the averaged quantities are then time dependent because the initial conditions $\mathbf{x}_0$ and $\mathbf{u}_{p,0}$ are imposed at $t = t_0$. More precisely, in a statistically steady flow, equation (4.1)

4 A methodology for assessment of particle-LES models

is equivalent to

$$\overline{f}(t|\mathbf{x}_0,\mathbf{u}_{p,0}) = \frac{1}{2T}\int_{-T}^{T} f\left(\mathbf{u}_{f@p}^{\tau_0}, \mathbf{u}_p^{\tau_0}, \mathbf{x}_p^{\tau_0}; t+\tau_0 | \mathbf{x}_0, \mathbf{u}_{p,0}\right)\, \mathrm{d}\tau_0 \qquad (4.2)$$

where T is some large time span over which f is averaged. $\mathbf{u}_{f@p}^{\tau_0}$, $\mathbf{u}_p^{\tau_0}$ and $\mathbf{x}_p^{\tau_0}$ denote fluid velocity seen by the particle, particle velocity and particle position respectively with initial conditions $\mathbf{x}_p^{\tau_0}(\tau_0) = \mathbf{x}_0$ and $\mathbf{u}_p^{\tau_0}(\tau_0) = \mathbf{u}_{p,0}$.

In forced homogeneous isotropic turbulence, which is the test configuration in the present work, equation (4.2) further simplifies. The homogeneity of the flow allows to set $\mathbf{x}_0 = 0$ without loss of generality. Furthermore, due to turbulence the particle's statistics become asymptotically independent of the particle's initial velocity.

Asymptotic independence of initial conditions is widely accepted for turbulent single phase flows (cf. e.g. Tennekes & Lumley, 1972; Townsend, 1975) although for reasons of completeness it should be noted that this premise might not apply in general (see George, 1989; Johansson et al., 2003). However, also recent investigations of turbulent flow are conducted under this assumption, likewise the present work.

Concerning particle-laden forced isotropic turbulence, for example Février et al. (2005) confirmed that the effect of initial conditions decays. In particular they point out that if t is larger than the integral time scale of the carrier flow and larger than a time scale which depends on the particle relaxation time, then particle statistics become independent of the initial conditions. Thus, for sufficiently large t, equation (4.2) simplifies in forced isotropic turbulence to

$$\overline{f}(t) = \frac{1}{2T}\int_{-T}^{T} f\left(\mathbf{u}_{f@p}, \mathbf{u}_p, \mathbf{x}_p; t+\tau_0\right)\, \mathrm{d}\tau_0. \qquad (4.3)$$

In forced isotropic turbulence all particle statistics are steady. Therefore a simple transform of variables yields

$$\overline{f} = \frac{1}{2T}\int_{-T}^{T} f\left(\mathbf{u}_{f@p}, \mathbf{u}_p, \mathbf{x}_p; \tau_0\right)\, \mathrm{d}\tau_0. \qquad (4.4)$$

This equation is simply the average of f along a particle's path. Thus, in isotropic turbulence the averaging operator $\overline{\cdot}$ reduces to Lagrangian averaging. Due to the independence of $\mathbf{x}_0$, this also corresponds to averaging over several particles. In the present work, both approaches are conducted for the numerical simulations. Statistics are computed by averaging over particles and time. It should be noted that this is not equivalent to spatial averaging because of preferential concentration.

4.2 Definition of assessment criteria

In the previous section the averaging operator was defined. This operator is used in the present section for the definition of assessment criteria. These criteria will serve in the following chapters for quantification of SGS effects on the particles and for assessment of particle-LES models.

For a specific application the assessment criteria are often self explanatory. For example, if one wants to predict the transport of air pollutants, then one will be interested in particle dispersion. If one wants to analyse the impact of particles colliding with a wall, then kinetic energy of the particles will be an interesting quantity. On the other hand, for chemical reactions, clustering of particles can be important. In this case, preferential concentration is an issue.

The present work is not aimed at one single specific application but follows a generalistic approach. The assessment criteria used here is based on a statistical point of view. Within the Lagrangian framework, each particle is characterised by its position, its velocity and the fluid velocity seen by the particle. From a statistical point of view, these three quantities define random processes. In the present work, the corresponding stochastic moments are considered as assessment criteria. In addition, preferential concentration is taken into consideration. Altogether, the assessment criteria read

A1 average particle position $\overline{x_p}(t)$ (first moment in particle position),

A2 fluid velocity seen by the particles $\overline{u_{f@p}}(t)$ (first moment in velocity seen by particles),

A3 average particle velocity $\overline{u_p}(t)$ (first moment in particle velocity),

A4 particle dispersion $\overline{x_{p,i}x_{p,i}}(t)$ (second moment in particle position),

A5 kinetic energy of the fluid seen by the particles $\frac{1}{2}\overline{u_{f@p,i}u_{f@p,i}}(t)$ (second moment in velocity seen by particles),

A6 kinetic energy of the particles $\frac{1}{2}\overline{u_{p,i}u_{p,i}}(t)$ (second moment in particle velocity),

A7 accumulation Σ (preferential concentration),

A8 fractal dimension d_{pc} (preferential concentration).

These are the quantities of interest in the following chapters. In chapter 5, the effect of small scale turbulence on these quantities is analysed. In chapters 6 and 7, particle-LES models are assessed with respect to these quantities.

Transport equations

The focus of the present work is on small scale effects on particles and on particle-LES models. The analytical assessment is either directly based on the statistical moments or on transport equations for these moments. In a Lagrangian framework, a transport equation is simply an equation for the time derivative. For the moments listed above, one can deduce

4 A methodology for assessment of particle-LES models

the following transport equations:

$$\frac{\mathrm{d}\overline{x_{p,i}}}{\mathrm{d}t} = \overline{u_{p,i}} \tag{4.5a}$$

$$\frac{\mathrm{d}\overline{x_{p,i}x_{p,i}}}{\mathrm{d}t}(t) = 2\int_{-\infty}^{t} \overline{u_{p,i}(\tau)u_{p,i}(t)}\,\mathrm{d}\tau. \tag{4.5b}$$

Assessment of some average $\overline{f}$ is equivalent to assessment of its time derivatve $\frac{\mathrm{d}\overline{f}}{\mathrm{d}t}$ if the effect of initial conditions is neglected. It was already explained above that there is sufficient evidence for this assumption. Therefore in the following it will be assumed that assessment of $\overline{f}$ is equivalent to assessment of $\frac{\mathrm{d}\overline{f}}{\mathrm{d}t}$.

For the analytical assessment in section 6.3, linear Stokes drag is assumed,

$$\frac{\mathrm{d}\mathbf{u}_p}{\mathrm{d}t} = \frac{1}{\tau_p}\left(\mathbf{u}_{f@p} - \mathbf{u}_p\right). \tag{4.6}$$

In this case one can derive additionally

$$\frac{\mathrm{d}\overline{u_{p,i}}}{\mathrm{d}t} = \frac{\overline{u_{f@p,i}} - \overline{u_{p,i}}}{\tau_p} \tag{4.7a}$$

$$\overline{u_{p,i}(\tau)u_{p,i}(t)} = \frac{1}{\tau_p^2}\int_{-\infty}^{t}\int_{-\infty}^{\tau} \overline{u_{f@p,i}(t_1)u_{f@p,i}(t_2)}e^{\frac{t_1+t_2-t-\tau}{\tau_p}}\,\mathrm{d}t_1\,\mathrm{d}t_2. \tag{4.7b}$$

Equations (4.5a) and (4.7a) reflect that $\overline{\mathbf{u}_p}$ and $\overline{\mathbf{x}_p}$ are exact if and only if $\overline{\mathbf{u}_{f@p}}$ is exact. Therefore the analytical analysis of the first moments (section 6.3.2) will be reduced to the analysis of $\overline{\mathbf{u}_{f@p}}$.

Equation (4.5b) furthermore shows that

$\overline{u_{p,i}u_{p,i}}$ and $\overline{x_{p,i}x_{p,i}}$ are exact

if and only if $\overline{u_{f@p,i}(\tau)u_{f@p,i}(t)}$ is exact for all $\tau \leq t$.

Therefore the analytical analysis of the second moments (sections 6.3.3 and 7.3) will be reduced to the analysis of $\overline{u_{f@p,i}(\tau)u_{f@p,i}(t)}$.

4.3 A priori and a posteriori analysis

Sections 4.1 and 4.2 concerned the definition of quantities which reflect the effect of turbulence on particles. The present section presents numerical techniques to compute these quantities such that one can differentiate between small scale and large scale effects. These techniques are called 'a priori' and 'a posteriori' analysis.

A priori analysis stands for comparison of filtered DNS data against reference DNS data. A posteriori analysis stands for comparison of LES data against reference DNS data. Thus, results from the a posteriori analysis depend on the fluid-LES model, in contrast to the a priori analysis. The combination of both allows to differentiate between errors from the fluid-LES model and effects of unresolved scales on particles.

The idea is to advance step by step from the DNS solution via a filtered DNS solution to the LES solution. However, if one looks into the details, then one will find that actually these are more than two steps. In other words, it needs to be clarified in detail what is meant by a priori analysis.

Figure 4.1 shows all steps which were conducted in a priori analysis in chapters 5 to 7. Each column corresponds to one data set. The leftmost column is the reference DNS data. The other columns are explained in the following.

The first step in an a priori analysis is the choice of a filter. In the present work, always box filters were implemented. After the choice of a filter one can advance to the second column.

The second step is to decide whether to sample this filtered field on the original DNS grid or on a grid which is as coarse as an LES grid. The latter choice will increase the error from interpolation of fluid velocity on particle position. For the analysis of small scale effects on particles it is desirable to separate effects of filtering from effects of interpolation errors. Thus, chapter 5 follows the first choice, the filtered field is sampled on the DNS grid. A priori analysis in chapter 5 stays in the second column. On the other hand, in order to test the performance of particle-LES models, the increased interpolation error must be taken into account as well. Therefore chapters 6 and 7 follow the second choice, the filtered field is sampled on a coarser grid. A priori analysis in these chapters advances to the third, fourth and fifth column.

The third step for the a priori analysis is to decide whether to compute particle paths from the filtered or from the unfiltered DNS field. A good particle-LES model will recover a DNS particle path in a statistical sense. Chapter 5 contains an analysis of the SGS effects which a good particle-LES model should emulate. Therefore here it makes sense to compute particle paths from the unfiltered DNS field and to record statistics of the filtered field along these paths. Consequently, interpolation in the second column feeds from particle positions from the first column.

The particle-LES models ADM and SOI (cf. chapters 6 and 7) do not take explicitly into account that the particle paths in DNS and LES deviate. Therefore an assessment of these models along DNS particle paths is not in contradiction with the respective model assumptions. Such an approach is conducted in chapters 6 and 7. On the same basis the models are assessed by analytical means in sections 6.3 and 7.3. Consequently, columns three and four also feed from particle positions from the first column.

On the other hand, the stochastic models analysed in chapter 6 are explicitly based on particle paths computed from the modelled field, cf. Fede et al. (2006). Therefore the numerical assessment of these models must be performed along the particle paths described by the models themselves. Thus, column five feeds from particle positions of the same column.

The stochastic models intend to reconstruct the particle paths in a statistical sense. Consequently the analytical assessment of these models (section 6.3) is partially based on an assumption on statistical exactness of particle paths. These assumptions are detailed in the corresponding section.

4 A methodology for assessment of particle-LES models

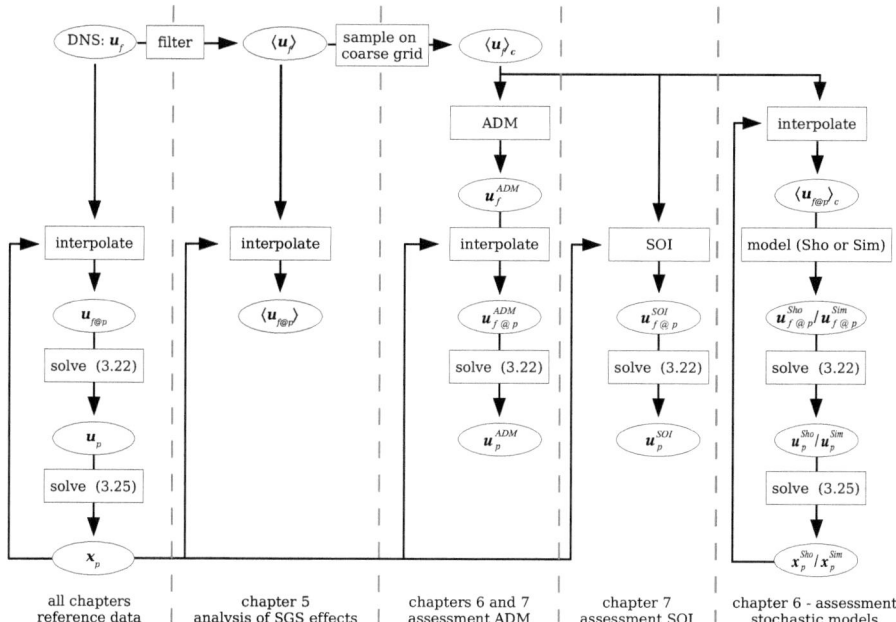

Figure 4.1: Schematic of steps in the a priori analysis. 'Sho' stands for the model proposed by Shotorban & Mashayek (2006), 'Sim' stands for the model proposed by Simonin et al. (1993), cf. chapter 6.

4.4 Conclusions of chapter 4

The present chapter consists of two parts. The first part (sections 4.1 and 4.2) focuses on quantities which reflect the effect of turbulence on particles. These quantities will be used in chapters 5 to 7 for the analysis of SGS effects on particles and for assessment of particle-LES models. The second part of this chapter (section 4.3) discusses several types of numerical experiments which are conducted in chapters 5 to 7 in order to separate SGS effects on these quantities from large scale effects.

In section 4.1 an averaging operator for particles in a turbulent flow is presented. This operator is applied throughout the present work. It defines statistical moments which serve as assessment criteria in the following chapters. In addition, preferential concentration is included as assessment criterion. In detail, the criteria read:

A1 average particle position $\overline{\mathbf{x}_p}(t)$ (first moment in particle position),

A2 fluid velocity seen by the particles $\overline{\mathbf{u}_{f@p}}(t)$ (first moment in velocity seen by particles),

A3 average particle velocity $\overline{\mathbf{u}_p}(t)$ (first moment in particle velocity),

A4 particle dispersion $\overline{x_{p,i}x_{p,i}}(t)$ (second moment in particle position),

A5 kinetic energy of the fluid seen by the particles $\frac{1}{2}\overline{u_{f@p,i}u_{f@p,i}}(t)$ (second moment in velocity seen by particles),

A6 kinetic energy of the particles $\frac{1}{2}\overline{u_{p,i}u_{p,i}}(t)$ (second moment in particle velocity),

A7 accumulation Σ (preferential concentration),

A8 fractal dimension d_{pc} (preferential concentration).

A motivation for these criteria is given in the present chapter. In addition, transport equations for the statistical moments A1, A3 and A4 are presented. In the following, either the statistical moments or the respective transport equations are computed. If a particle-LES model reconstructs the transport equation of one of the moments A1, A3 or A4, then it will be concluded that it predicts the respective moment correctly.

The second part of this chapter, section 4.3, discusses the techniques of a priori and a posteriori analysis. These are techniques for the analysis of small scale turbulence or for model assessment, implemented in the following chapters.

A posteriori analysis means comparison of DNS data against LES data. A priori analysis means comparison of DNS data against data from filtered DNS. Section 4.3 shows that for particle-laden flow several types of a priori analysis are possible. The purpose of the analysis specifies which type is recommendable. Section 4.3 explains which type is to be selected for which purpose.

5 A numerical study on requirements for a particle-LES model

The purpose of computing is insight not numbers.
C. Hastings

In Large Eddy Simulation it is common practice to neglect the effect of the subgrid scales (SGS) on the particles although for most configurations it is not clear to which extent the small scales affect the particles' dynamics. In the present chapter it is shown by numerical experiments that actually this effect is not negligible. In particular, it is shown that neglection leads to underestimation of kinetic energy and overestimation of integral time scales. A particle-LES model is required to compensate for this.

Particles with very high relaxation time τ_p (i.e. high inertia) are not very sensitive to high frequency fluctuations. Therefore it is commonly assumed that for high τ_p the effect of the unresolved scales can be safely neglected, cf. e.g. Yamamoto *et al.* (2001). The present chapter shows that this holds for the kinetic energy of the particles but not for the rate of dispersion.

In particular, the present chapter shows that SGS effects on the particles depend *qualitatively* on Reynolds number. This could only be detected by simulations at very high Reynolds number. To date, the present study is the only study of SGS effects on particles at high Reynolds number. Furthermore, there are no other studies at a comparable range and resolution in Stokes numbers.

With this data, it was for the first time possible to formulate scaling laws for the Stokes number dependence of kinetic energy and integral time scale seen by the particles. These scaling laws are important for understanding the physical effects involved. However, they are only a side product of this study. The main focus is on the effects which need to be reconstructed by a particle-LES model.

This chapter is organised as follows. Section 5.1 contains an overview of published works which are strongly related to the present chapter. In sections 5.2 and 5.3, the parameters for the present test configurations are detailed. Section 5.4 presents results for validation of the methods. Section 5.5 contains the core of this chapter, namely results of an a priori and an a posteriori analysis of the effect of small scale turbulence on particles.

5.1 Relation to previous works

One of the first studies concerning SGS effects on particles was conducted by Armenio, Piomelli & Fiorotto (1999). They analysed inertia free particles, i.e. $\tau_p = 0$, and found that SGS turbulence has a significant effect on the particles. For $0 < \tau_p < \infty$, there is little work in this direction. The only two widely referenced works in this field are the analyses by Kuerten & Vreman (2005), Yamamoto et al. (2001) and Fede & Simonin (2006).

Kuerten and Vreman (see also Kuerten, 2008) studied particles in a turbulent channel flow. They found that LES underpredicts the movement of particles towards the wall, i.e., the spatial distribution of particles is not predicted correctly by LES.

Yamamoto et al. conducted an a priori analysis in turbulent channel flow, i.e., they compared DNS data against filtered DNS data. The authors characterised the effect of small scale turbulence by comparing particle trajectories computed from filtered DNS against trajectories from unfiltered DNS. The Reynolds number of the simulation was very high (Reynolds number based on friction velocity was $Re_\tau = 644$) but the authors themselves state that, due to limited computational capacity, they could not resolve all scales present in the flow.

Fede and Simonin conducted an a priori analysis in isotropic turbulence. In contrast to Yamamoto et al. they resolved all scales present in the flow. To achieve this, they restricted themselves to a relatively low Reynolds number of $Re_\lambda = 34$. At this Reynolds number, the energy spectrum of the flow shows no inertial subrange. Thus it is not clear whether their results can be applied to high Reynolds number flows. Therefore more investigations at high Reynolds number are needed.

The present chapter contains results of such investigations. It extends and complements the findings of Fede and Simonin. Besides the different Stokes and Reynolds numbers, the focus is also somewhat different. Fede and Simonin focused on effects of SGS turbulence on particle collision. The present chapter considers configurations where collision is negligible and focuses on the criteria from section 4.2 at high Reynolds number and a broad range of Stokes numbers. Particle statistics from simulations of isotropic turbulence up to Reynolds number $Re_\lambda = 99$ are presented. At this Reynolds number, the energy spectrum of the carrier fluid clearly shows an inertial subrange, i.e., the Reynolds number is sufficiently high such that Kolmogorov's hypotheses are applicable.

The methodology from chapter 4 is applied in order to quantify SGS effects on particles. This analysis gives insight into the mechanisms which a particle LES model must emulate.

Thus, this chapter addresses the two questions "What happens if subgrid scale effects are neglected?" and "Which properties must a particle-SGS model have?".

5.2 Numerical Simulation of the carrier flow

As mentioned above, previously published results were confined to low Reynolds number or to underresolved DNS. The simulations presented in this chapter do not show any of these deficiencies. This is shown in the following. In addition, spectra from DNS, filtered DNS and LES are presented. The implemented simulation techniques and numerical methods were already described in chapter 3.

The flow was computed at four Reynolds numbers, namely $Re_\lambda = 34$, $Re_\lambda = 52$, $Re_\lambda = 99$ and $Re_\lambda = 265$. The Reynolds number $Re_\lambda = \frac{\lambda u_{rms}}{\nu}$ is based on the transverse Taylor mi-

5 A numerical study on requirements for a particle-LES model

croscale λ and the rms value of one (arbitrary) component of the fluctuations u_{rms}. The highest Reynolds number serves not for SGS analysis but only for code validation, see below.

In all computations the flow was solved in a cube on a staggered Cartesian equidistant grid. For $Re_\lambda = 34$, 52 and 99, the size of the computational box and the cell width was chosen in dependence of the Reynolds number such that all scales are resolved, based on the criteria stated by Pope (2000). They read (cf. section 3.1.1):

C1 The DNS should resolve $\kappa \leq 1.5/\eta_K$, i.e., Δx should be smaller than $\eta_K \pi/1.5 \approx 2.1 \eta_K$. Table 5.1 shows that this requirement is readily fulfilled.

C2 The computational box size L should be larger than 8 integral length scales L_f. Table 5.1 shows that this requirement is fulfilled for $Re_\lambda = 34$, 52 and 99 but not for $Re_\lambda = 265$ if L_f is computed from the model spectrum of section 2.1.3.

Table 5.1: Simulation parameters and Eulerian statistics from DNS of forced isotropic turbulence. Reynolds number Re_λ, number of grid points N, length of computational box L, cell width Δx, range of forced wavenumbers $[\kappa_0, \kappa_1]$, rate of dissipation ϵ, Kolmogorov length scale η_K, Kolmogorov time scale τ_K, integral length scale L_f, time scale of energy containing eddies k_f/ϵ, filter width Δ and ratio of kinetic energy of the filtered field $\langle k_f \rangle$ to kinetic energy of the unfiltered field k_f. Model spectrum from section 2.1.3.

	DNS			
Re_λ	34	52	99	265
N	128^3	256^3	512^3	1030^3
L/λ	16.3	23.8	39.9	48.3
L/L_f from simulation	9.77	11.9	12.2	13.4
L/L_f from model spectrum	8.0	9.15	9.9	5.75
$\Delta x/\lambda$	0.127	0.093	0.078	0.047
$\Delta x/\eta_K$	1.47	1.34	1.54	1.54
$[\kappa_0, \kappa_1]\lambda$	[0.996, 1.49]	[0.514, 1.54]	[0.33, 0.881]	[0.077, 0.33]
$\epsilon \lambda^2/u_{rms}^2/\nu$	15.1	14.99	15.8	14.8
η_K/λ	0.087	0.070	0.050	0.030
$\tau_K u_{rms}/\lambda$	0.257	0.248	0.252	0.28
L_f/λ from simulation	1.67	2.00	3.27	3.6
L_f/λ from model spectrum	2.04	2.6	4.03	8.4
$k_f/\epsilon\, u_{rms}/\lambda$	3.40	5.15	9.38	28.5
$\Delta/\Delta x$	5	7	9	n.a.
$\langle k_f \rangle / k_f$	88%	87%	88%	n.a.

In the present chapter, the testcase $Re_\lambda = 265$ serves only for code validation, section 5.4. At this Reynolds number, requirement C2 was not fulfilled due to computational limitations. This means that for this Reynolds number the box size is too small to allow for reliable statistics. Therefore the analysis of small scale turbulence is restricted to the smaller Reynolds numbers.

In chapter 7, results at $Re_\lambda = 265$ will be presented for the new model. It should be noted that due to the too small box size these results have only an indicative character.

One crucial point for the simulation of isotropic turbulence is that the reference length scales η_K and L_f are a result of the simulation, i.e., if the box is too small or the grid too coarse then these quantities are not computed correctly. This complicates validation of the requirements stated above. One way to check that η_K is computed correctly is to compare the numerically computed values for η_K/λ against their theoretical counterparts. The numerically computed values are listed in table 5.1. The theoretical values can be derived from equation (2.14),

$$\frac{\eta_K}{\lambda} = \frac{1}{15^{1/4}\sqrt{Re_\lambda}}. \tag{5.1}$$

The numerically computed values and the values computed from this formula are identical for all three Reynolds numbers, indicating that requirement C1 is indeed fulfilled. Of course, this check is only a necessary and not a sufficient condition for requirement C1 because λ and η_K are both computed from the simulation results.

Concerning requirement C2, there is no analytical equation for L_f/λ. The only workaround is to compute L_f/λ from the model spectrum of section 2.1.3. These values are listed in table 5.1. They are larger than the numerically computed values for L_f/λ (see same table) indicating that the large scales are not well resolved. Considering that in the present work the focus is on small scales, the deviations are nevertheless acceptable expect for the testcase at $Re_\lambda = 265$. Therefore this testcase was not used for analysis of small scale statistics. It should be noted that for $Re_\lambda = 34$, 52 and 99, requirement C2 is fulfilled, independent whether L_f/λ is computed from the model spectrum or from the simulation results.

The energy spectrum functions from DNS are plotted in figure 5.1. The simulation at $Re_\lambda = 99$ shows a well established inertial subrange with $E(\kappa) \sim \kappa^{-5/3}$. This is a prerequisite for investigating effects at high Reynolds number turbulence.

Figure 5.2 shows a comparison of the DNS spectra against the model spectrum from section 2.1.3. Here, scaling on the wavenumber axis is based on the Kolmogorov length scale η_K. At low wavenumbers, the spectra show deviations, explaining the deviations in L_f/λ. Actually the low wavenumber regime is strongly dependent of the forcing scheme. In the present work, this regime is of minor interest because the focus is on small scale effects. In LES and DNS the same forcing scheme was applied and therefore the deviations to the model spectrum are not an issue.

At all Reynolds numbers, the model spectra show a slightly extended dissipative range in comparison to the DNS spectrum, i.e., the decay at high wavenumbers is overpredicted in DNS or underpredicted in the model spectrum. However, for the purpose of the present work the match between DNS and model spectrum is satisfactory.

The small scale effects on the particles are quantified by a priori and a posteriori analysis, i.e., by comparison of DNS data against filtered DNS data and LES data respectively.

In the a priori analysis, small scale turbulence is assessed along the DNS particle trajectories, following the reasoning of section 4.3. This means that the particle transport equation (2.32) is solved using the (unfiltered) fluid velocity $\mathbf{u}_{f@p}$. For computing statistical quantities, the large scale fluctuations were extracted using a box filter (cf. section 3.1.2),

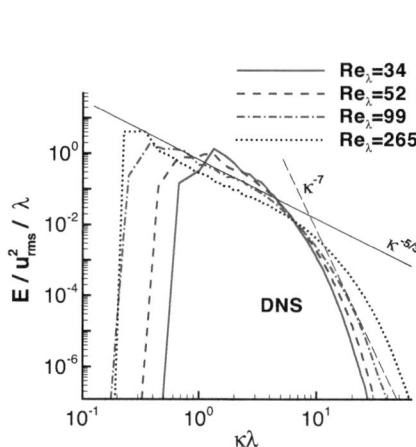

Figure 5.1: Instantaneous energy spectrum functions from DNS together with a line proportional to $\kappa^{-5/3}$ (continuous straight line) and a line proportional to κ^{-7} (long dashed straight line).

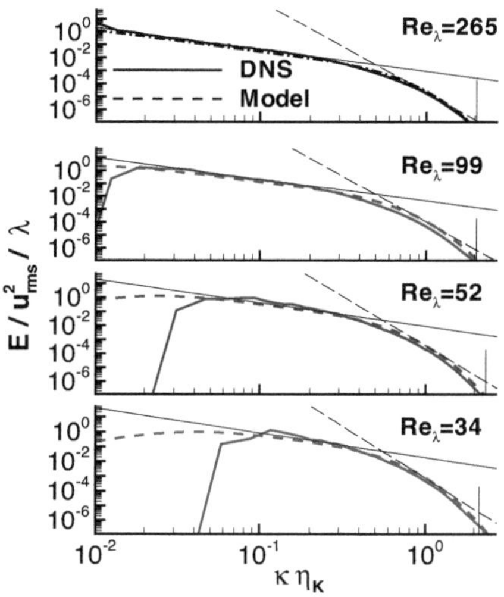

Figure 5.2: Instantaneous energy spectrum functions from DNS and Pope's model (cf. section 2.1.3) together with lines proportional to $\kappa^{-5/3}$ (continuous straight lines) and lines proportional to κ^{-7} (long dashed stragiht lines). The vertical lines on the right hand side mark the DNS cutoff wavenumber $\kappa = \pi/\Delta x$.

$$\langle \mathbf{u}_f \rangle (\mathbf{x}, t) = \frac{1}{\Delta^3} \iiint_{[-\Delta/2, \Delta/2]^3} \mathbf{u}_f(\mathbf{x} + \mathbf{r}, t) \, d\mathbf{r}. \tag{5.2}$$

$\mathbf{u}_f$ is the fluid velocity computed from DNS and Δ is the filter width. Δ is set such that the energy of the filtered field $\langle k_f \rangle = \overline{\langle u_{f,i} \rangle^2}/2$ is 87 - 88 % of the energy of the unfiltered field $k_f = \overline{u_{f,i}^2}/2$, cf. table 5.1. As before, $\overline{\cdot}$ denotes spatial and temporal averaging. The reader is reminded that for the testcase $Re_\lambda = 265$ no small scale statistics were recorded and consequently this DNS was not filtered.

$\langle \mathbf{u}_f \rangle$ does not contain the high wavenumber fluctuations. In the a priori analysis, the effect of the small scales is analysed by comparing statistics of $\mathbf{u}_f$ with those of $\langle \mathbf{u}_f \rangle$ on the particles' positions.

For the a posteriori analysis, the Lagrangian dynamic Smagorinsky model proposed by Meneveau et al. (1996) (cf. section 3.1.2) was implemented. For each Reynolds number, the

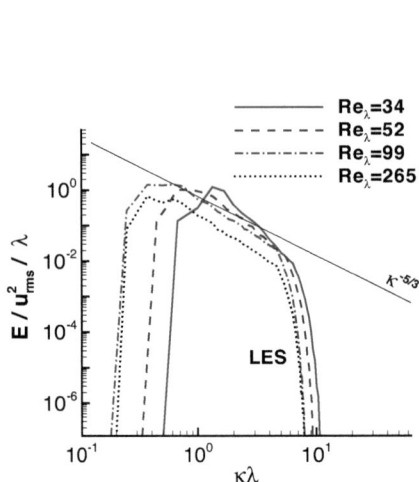
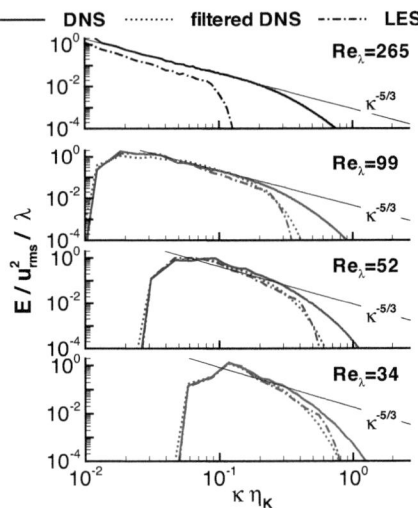

Figure 5.3: Instantaneous energy spectrum functions from LES together with a line proportional to $\kappa^{-5/3}$ (continuous straight line).

Figure 5.4: Instantaneous energy spectrum functions from DNS, filtered DNS and LES together with lines proportional to $\kappa^{-5/3}$ (continuous straight lines).

resolution of the LES was chosen such that the kinetic energy resolved by LES $[k_f]$ and the kinetic energy of the filtered DNS field $\langle k_f \rangle$ are approximately equal. The LES velocity is denoted by $[\mathbf{u}_f]$. Parameters for the LES can be found in table 5.2. $[\epsilon]$ denotes the resolved dissipation rate

$$[\epsilon] = 2(\nu + \nu_t)\overline{[S_{ij}][S_{ij}]}, \qquad [S_{ij}] = \frac{1}{2}\left(\frac{\partial [u_{f,i}]}{\partial x_j} + \frac{\partial [u_{f,j}]}{\partial x_i}\right). \tag{5.3}$$

It should be noted that one inherent problem in LES of isotropic turbulence is the computation of the Reynolds number Re_λ because rms velocity and Taylor length scale depend on small scale turbulence. Therefore, in the present work, for each Reynolds number first a DNS was conducted. The range of forced wavenumbers $[\kappa_0, \kappa_1]$ and the energy contained in that range was set such that the desired Reynolds number is attained. Then, exactly these parameters were used to conduct the corresponding LES, i.e., the energy contained in the wavenumber range $[\kappa_0, \kappa_1]$ is equal in LES and DNS.

Instantaneous energy spectra of the LES are plotted in figure 5.3. It should be noted that the cutoff wavenumber based on λ is approximately equal at all Reynolds numbers. In figure 5.4, the spectra from DNS, filtered DNS and LES can be compared easily. Evidently the match between filtered DNS and LES spectra is very good, which is a pre-

5 A numerical study on requirements for a particle-LES model

requisite for a quantitative comparison of results from a priori and a posteriori analysis.

	LES			
Re_λ	34	52	99	265
N	32^3	42^3	64^3	42^3
$\Delta x/\lambda$	0.509	0.567	0.623	1.15
$[k_f]/[\epsilon]\, u_{rms}/\lambda$	3.5	5.9	11.1	32.76
$[k_f]/k_f$	87%	86%	87%	92%

Table 5.2: Parameters for LES of forced isotropic turbulence and time scale of energy containing eddies $[k_f]/[\epsilon]$ computed from resolved scales.

5.3 Parameters for the discrete particle simulations

The previous section contained the parameters for the simulations of the carrier flow. The present section contains parameters for the simulation of the particles.

In all computations the methods described in section 3.3 were implemented. The density of the particles was set to $\rho_p = 1800\rho$ where ρ is the density of the fluid. In each simulation the particles were divided in 24 fractions with different diametre d. The maximum diametre was chosen in dependence of the Reynolds number in such a way that the diametre of the largest particles equals the Kolmogorov length scale. Thus, the particles can be treated as point particles, cf. section 2.2.1.

In LES, the SGS effect on the particles was not modelled because this chapter aims at an analysis of these effects. Therefore, in LES the equation

$$\frac{d[\mathbf{u}_p]}{dt} = -\frac{c_D Re_p}{24\tau_p}([\mathbf{u}_p] - [\mathbf{u}_{f@p}]) \tag{5.4}$$

was solved with the same method as in DNS.

For single particle statistics (sections 5.5.1 to 5.5.3), in all simulations 24 fractions of particles were traced with 80000 particles per fraction. For the analysis of preferential concentration (section 5.5.4), 8 fractions of particles were traced. Here, the number of particles in the simulations was set in dependence of the Reynolds number (cf. section 5.5.4).

In all simulations the particles were initialised at random positions (homogeneous distribution) inside the computational box and traced until a statistical steady state was obtained. Then, 1000 time records were taken within a time span of $T = 250\lambda/u_{rms}$ for computing statistics. The temporal resolution of the statistics equals approximately the Kolmogorov time scale. With this temporal resolution, the Lagrangian correlation functions could be resolved for all Stokes numbers. The time span was large enough to guarantee that averaging in time cancels out oscillations caused by the forcing scheme.

Also in terms of particle time scales, T is large enough to guarantee reliable statistics. With $\epsilon = 15\nu u_{rms}^2/\lambda^2$ (cf. section 2.1.3), T/τ_p can be written as $T/\tau_p = 250\sqrt{15}/St \approx 968.2/St$. In all simulations, $St \leq 100$, thus $T/\tau_p \geq 9.68$. Hence, statistics were sampled over at least 9.68 times the particle relaxation time.

5.4 Validation

The code was validated via probability density functions (PDFs) for the particle acceleration. To this end, data from the DNS at $Re_\lambda = 265$ was compared with data from a DNS conducted by Biferale et al. (2004) and an experiment conducted by Ayyalasomayajula et al. (2006).

Biferale et al. conducted a DNS at $Re_\lambda = 280$ with inertia free particles (i.e. $St = 0$). Ayyalasomayajula et al.'s experiment was at $Re_\lambda = 250$ with particle Stokes numbers $St = 0.09 \pm 0.03$. Correspondingly, in the present simulation two particle fractions were traced, one at $St = 0$ and another at $St = 0.1$. Each fraction consists of 960000 particles. Figure 5.5 shows that the results from the present simulations agree very well with the referenced data.

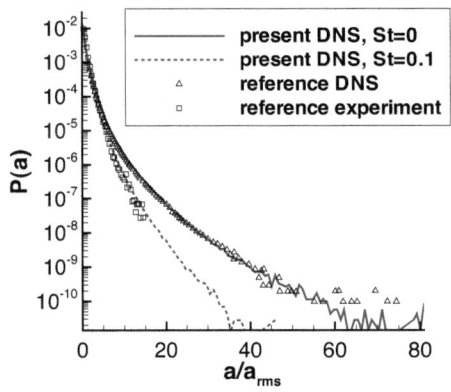

Figure 5.5: Probability density function of particle acceleration **a**. X-axis is normalised with respect to the rms value of **a**. Triangles: reference DNS of $St = 0$ particles conducted by Biferale et al. (2004). Squares: reference experiment of $St = 0.09 \pm 0.03$ particles conducted by Ayyalasomayajula et al. (2006) (renormalised).

5.5 Effect of the SGS turbulence on the particles

The present section contains the core of this chapter, namely the analysis of SGS effects on the particles. As mentioned above, the testcase $Re_\lambda = 265$ was excluded due to the small box size. The statistical moments listed in chapter 4 define the criteria for this analysis.

First, kinetic energy and integral time scales are analysed. Then, dispersion and preferential concentration follow.

5.5.1 Kinetic energy

In the following, SGS effects on kinetic energy seen by the particles and on kinetic energy of the particles are presented. In particular,

- scaling laws for the kinetic energy of the fluid seen by the particles are formulated,

5 A numerical study on requirements for a particle-LES model

- it is shown that the SGS kinetic energy seen by the particles depends strongly on Stokes number,
- and it is shown that the kinetic energy of the particles is underpredicted in LES.

Scaling laws for the kinetic energy of the fluid seen by the particles

The first quantity under consideration is the kinetic energy seen by the particles. This quantity depends on Stokes number due to clustering. In the following it is shown that the analysed data indicates the existence of two universal subranges of Stokes numbers where the Stokes number dependence can be described by scaling laws.

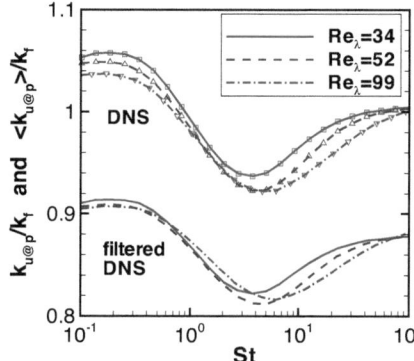

Figure 5.6: A priori analysis: Unfiltered and filtered kinetic energy of the fluid seen by the particles computed from DNS.

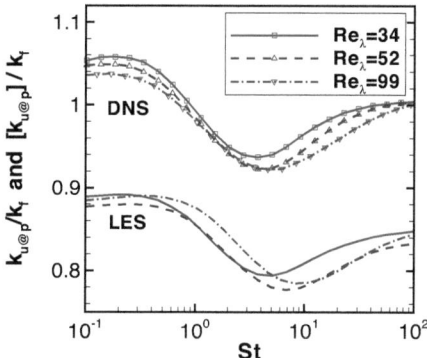

Figure 5.7: A posteriori analysis: Kinetic energy seen by the particles in DNS and LES.

In figure 5.6, the unfiltered and filtered kinetic energy of the fluid seen by the particles

$$k_{u@p} = \frac{1}{2}\overline{u_{f@p,i}^2} = \frac{1}{2}\overline{u_{f,i}^2(\mathbf{x}_p(t),t)} \quad \text{and} \quad \langle k_{u@p} \rangle = \frac{1}{2}\overline{\langle u_{f@p,i} \rangle^2} = \frac{1}{2}\overline{\langle u_{f,i} \rangle^2(\mathbf{x}_p(t),t)} \quad (5.5)$$

computed from DNS are depicted. In figure 5.7, the corresponding result from LES is shown,

$$[k_{u@p}] = \frac{1}{2}\overline{[u_{f@p,i}]^2} = \frac{1}{2}\overline{[u_{f,i}]^2(\mathbf{x}_p(t),t)}. \quad (5.6)$$

Figure 5.6 shows that filtering leads to a decrease in the kinetic energy seen by the particles, as expected. Both, filtered and unfiltered kinetic energy, show clear Stokes number dependence due to clustering. Particles with Stokes numbers smaller than one tend to cluster in

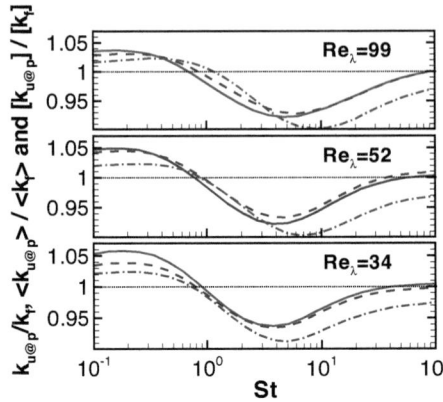

Figure 5.8: A priori and a posteriori analysis: Kinetic energy of the fluid seen by the particles, scaled by resolved kinetic energy. Continuous lines: DNS, dashed lines: filtered DNS, dash-dotted lines: LES.

regions where kinetic energy is higher than average, particles with Stokes numbers greater than one tend to cluster in regions where kinetic energy is lower than average. Reynolds number dependence will be discussed below.

First, consider the LES result, figure 5.7. On first sight it seems that the LES result is equal to the filtered DNS result. For a more detailed comparison, both results are plotted on top of each other in figure 5.8. In that figure, the filtered and LES kinetic energy seen by the particles is rescaled to $\langle k_f \rangle$ and $[k_f]$, respectively. Now one can clearly observe a shift along the St-axis between LES and DNS. This leads to the hypothesis that if the St-axis is rescaled by the correct characteristic time scale, then a match between DNS and LES results can be obtained.

In the present work, a good match was achieved by scaling with the resolved Kolmogorov time scale for low particle relaxation times and with the resolved eddy decay time for high particle relaxation times. In the following, this hypothesis is first formulated in terms of two scaling laws. Then, numerical data is provided supporting these laws.

Scaling Law 1. *Scaling of kinetic energy seen by particles at small relaxation time.*
Define the Stokes number St_S based on the smallest resolved scales, i.e.,

$$\text{in DNS set } St_S = St = \frac{\tau_p}{\tau_K} = \frac{\tau_p \sqrt{\epsilon}}{\sqrt{\nu}}, \quad \text{in LES set } St_S = \frac{\tau_p \sqrt{[\epsilon]}}{\sqrt{\nu + \nu_t}}. \tag{5.7}$$

Then, $k_{u@p}/k_f$, $\langle k_{u@p} \rangle / \langle k_f \rangle$ and $[k_{u@p}]/[k_f]$ scale with $\log St_S$ around $St_S = 1$. The scaling factor is approximately -0.14,

$$\left. \begin{array}{l} \frac{k_{u@p}}{k_f} \sim -0.14 \log \frac{\tau_p \sqrt{\epsilon}}{\sqrt{\nu}}, \\ \frac{\langle k_{u@p} \rangle}{\langle k_f \rangle} \sim -0.14 \log \frac{\tau_p \sqrt{\epsilon}}{\sqrt{\nu}}, \\ \frac{[k_{u@p}]}{[k_f]} \sim -0.14 \log \frac{\tau_p \sqrt{[\epsilon]}}{\sqrt{\nu+\nu_t}} \end{array} \right\} \text{ around } St_S \approx 1. \tag{5.8}$$

5 A numerical study on requirements for a particle-LES model

Scaling Law 2. *Scaling of kinetic energy seen by particles at high relaxation time. Define the Stokes number St_L based on the large eddy decay time scale, i.e.,*

$$\text{in DNS set } St_L = \frac{\tau_p \epsilon}{k_f}, \qquad \text{in LES set } St_L = \frac{\tau_p [\epsilon]}{[k_f]}. \tag{5.9}$$

Then, $k_{u@p}/k_f$, $\langle k_{u@p}\rangle / \langle k_f\rangle$ and $[k_{u@p}]/[k_f]$ scale with $\log St_L$ around $St_L = 0.9$. The scaling factor is approximately 0.9,

$$\left.\begin{array}{l}\frac{k_{u@p}}{k_f} \sim 0.9 \log \frac{\tau_p \epsilon}{k_f}, \\ \frac{\langle k_{u@p}\rangle}{\langle k_f\rangle} \sim 0.9 \log \frac{\tau_p \epsilon}{k_f}, \\ \frac{[k_{u@p}]}{[k_f]} \sim 0.9 \log \frac{\tau_p [\epsilon]}{[k_f]}\end{array}\right\} \text{ around } St_L \approx 0.9. \tag{5.10}$$

The numerical results support both scaling laws. In figure 5.9, DNS and filtered DNS data is plotted with the scaling of law 1. Around $St_S \approx 1$, all data collapse into a linear function in accordance with the scaling law. In figure 5.10, the corresponding LES data is plotted. Here, only the DNS result from the highest Reynolds number is included for reasons of clarity. Again, the data is in accordance with scaling law 1.

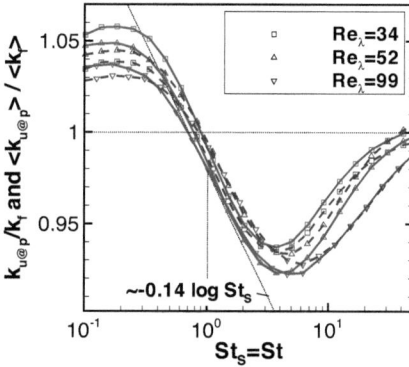

Figure 5.9: A priori analysis: Kinetic energy of the fluid seen by the particles computed from DNS, scaled by resolved kinetic energy. Continuous lines: DNS, dashed lines: filtered DNS.

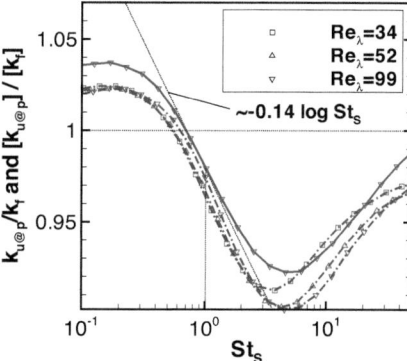

Figure 5.10: A posteriori analysis: Kinetic energy of the fluid seen by the particles, scaled by resolved kinetic energy and smallest resolved time scale. Continuous line: DNS at $Re_\lambda = 99$, dash-dotted lines: LES.

For validation of scaling law 2, data from DNS and filtered DNS is shown in figure 5.11, scaled by St_L. Again, the data shows good agreement with the scaling law. Figure 5.12

shows the corresponding LES result. Apparently no good match between DNS and LES data can be observed. The deviations are actually due to the interpolation error in LES. In contrast to DNS, the LES grid is coarse enough to show significant interpolation error. The implemented fourth-order scheme does not preserve kinetic energy, i.e., the kinetic energy of the interpolated data is smaller than the kinetic energy of the original data. Therefore, the results from figure 5.12 were compared to results obtained from LES with second-order conservative interpolation, cf. figure 5.13. The latter scheme is described in more detail in Gobert et al. (2006) and analysed in chapter 7. It is not a standard interpolation scheme and in the present case it leads to higher kinetic energy than the fourth-order interpolation scheme. With this scheme, the scaling proposed in law 2 leads to good agreement between DNS and LES data.

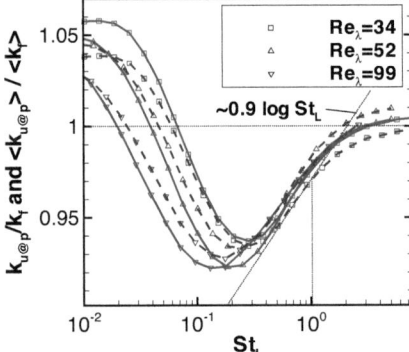

Figure 5.11: A priori analysis: Kinetic energy of the fluid seen by the particles, scaled by resolved kinetic energy and large eddy decay time. Continuous lines: DNS, dashed lines: filtered DNS.

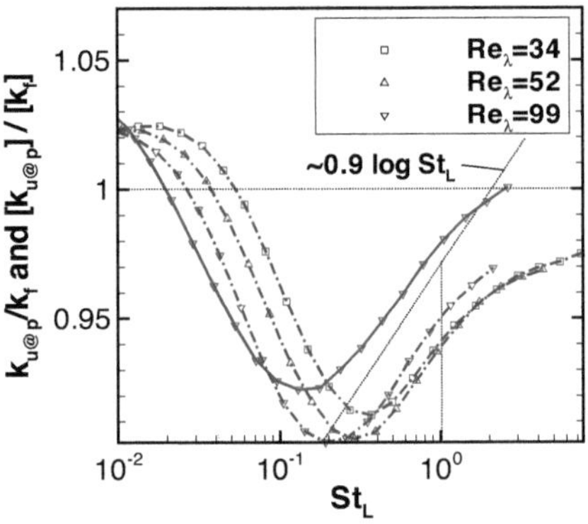

Figure 5.12: A posteriori analysis: Kinetic energy of the fluid seen by the particles, scaled by resolved kinetic energy and resolved large eddy decay time. Continuous line: DNS at $Re_\lambda = 99$, dash-dotted lines: LES with 4th-order interpolation.

However, the second-order interpolation leads to significant overprediction of kinetic energy, clearly visible in figure 5.13 at $Re_\lambda = 34$ and $St_L > 2.5$. Therefore in the following no more results from this scheme are discussed.

Concluding, the numerical data supports the two scaling laws. In summary, they state that the range of particle relaxation times can be divided in two regimes, namely $St_S \approx 1$ and $St_L \approx 0.9$. In these regimes, particle dynamics are governed by high frequency and low frequency fluctuations, respectively. In the first regime, increasing particle relaxation time means decreasing kinetic energy seen by the particles, in the second regime increasing relaxation time means increasing kinetic energy. The relaxation time where kinetic energy seen by the particles attains its minimum is in between these two regimes.

A particle-LES model should recover the regime around $St_S \approx 1$ where particle dynamics

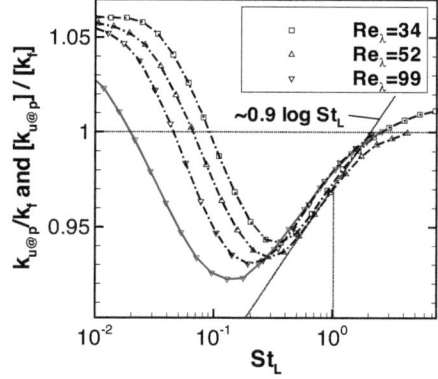

Figure 5.13: A posteriori analysis: Kinetic energy of the fluid seen by the particles, scaled by resolved kinetic energy and resolved large eddy decay time. Continuous line: DNS at $Re_\lambda = 99$, dash-dotted lines: LES with 2nd-order interpolation.

are governed by short living eddies. Concerning the second regime, a particle-LES model would have to modify the large time scales in order to recover the DNS result. This is in contrast to the idea of a particle-LES model. If for some application this regime is of absolute importance, then a proper choice of a fluid-LES model can help out. Equation (2.14) provides a link between ν_t and the large eddy decay time. It holds

$$\frac{[k_f]}{[\epsilon]} = \frac{[\lambda]^2}{10(\nu + \nu_t)} \quad \text{and} \quad \frac{k_f}{\epsilon} = \frac{\lambda^2}{10\nu}. \tag{5.11}$$

Thus,

$$\frac{[k_f]}{[\epsilon]} = \frac{k_f}{\epsilon} \quad \text{if and only if} \quad \nu_t = \left(\frac{[\lambda]^2}{\lambda^2} - 1\right)\nu. \tag{5.12}$$

If the fluid-LES model provides ν_t such that (5.12) holds, then the regime around St_L should be recovered by LES.

Concerning Reynolds number dependence, the two scaling laws mean that at high Reynolds number, where a large bandwidth of length and time scales is observable, the two regimes are well separated. This again means that, at sufficiently high Reynolds number, a subrange in between these two regimes must exist where the kinetic energy seen by the particles follows a universal law. A particle-LES model must be capable to recover that subrange. However, to this end reference data from very high Reynolds number is necessary. Currently computational capabilities are not sufficient to generate this reference data and state of the art experimental techniques are not capable to produce data at the required accuracy. Therefore the existence of this subrange remains a hypothesis for now.

Furthermore, figures 5.6 and 5.8 show that the shift in $k_{u@p}$ caused by filtering is essentially independent of Stokes number,

$$\frac{\langle k_{u@p}\rangle(St)}{\langle k_{u@p}\rangle(0)} \approx \frac{k_{u@p}(St)}{k_{u@p}(0)}. \tag{5.13}$$

The reader is reminded that in filtered and unfiltered DNS, the particles were traced along the same paths. Thus, a comparison of filtered and unfiltered results gives no indication on the scales which drive clustering but only on the scales which determine locations for clustering. If these locations were determined by small scales only, then filtering would obscure the Stokes number dependence. Figures 5.6 and 5.8 show that this is untrue and therefore mainly the large scales determine locations for clustering. Thus, a particle-LES model must retain locations for clustering.

Stokes number dependence of SGS kinetic energy seen by the particles

The stochastic particle-LES models of Shotorban & Mashayek (2006) and Simonin *et al.* (1993) (cf. chapter 6) use the SGS kinetic energy seen by the particles as model parameter. They assume that this parameter is independent of Stokes number. In the following it is shown that this assumption is questionable.

First, decompose $\mathbf{u}_{f@p}$ in large scale velocity $\langle \mathbf{u}_{f@p} \rangle$ and small scale velocity $\mathbf{u}'_{f@p}$, computed from $\mathbf{u}'_{f@p} = \mathbf{u}_{f@p} - \langle \mathbf{u}_{f@p} \rangle$. The kinetic energy can be decomposed by

$$k_{u@p} = \frac{1}{2}\overline{u^2_{f@p,i}} = \langle k_{u@p} \rangle + \overline{\langle u_{f@p,i} \rangle u'_{f@p,i}} + \underbrace{\frac{1}{2}\overline{u'_{f@p,i} u'_{f@p,i}}}_{=k'_{u@p}} . \qquad (5.14)$$

$\overline{\langle u_{f@p,i} \rangle u'_{f@p,i}}$ is the covariance of large and small scales. It should be noted that for sharp spectral filters, $\overline{\langle u_{f@p,i} \rangle u'_{f@p,i}} = 0$ but for most other filters, including Smagorinsky and top hat filter, this does not hold. This can be readily verified by writing $\langle u_{f,i} \rangle u'_{f,i}$ in terms of the filter transfer function $\mathcal{FT}(G)(\kappa)$ and the Fourier transformed velocity:

$$\langle \mathbf{u}_f \rangle . \mathbf{u}'_f = \iiint \mathcal{FT}(G)(\kappa) \, (1 - \mathcal{FT}(G)(\kappa)) \, \|\mathcal{FT}(\mathbf{u}_f)(\kappa)\|^2 \, d\kappa \qquad (5.15)$$

Apparently, $\mathcal{FT}(G)(\kappa) \, (1 - \mathcal{FT}(G)(\kappa))$ is not zero in general.

In figures 5.14 and 5.15, $\overline{\langle u_{f@p,i} \rangle u'_{f@p,i}}$ and $k'_{u@p}$ are depicted. At $Re_\lambda = 34$ most of the residual turbulent kinetic energy is in the covariance between large and small scales. Due to the limited spectral range, the filter cannot clearly separate the small scales from the large scales. With increasing Reynolds number, the magnitude of the covariance $\overline{\langle u_{f@p,i} \rangle u'_{f@p,i}}$ decreases whereas the magnitude of $k'_{u@p}$ increases. This is due to the broadening of the energy spectrum and increase of filter width with respect to the Kolmogorov length scale. For LES at very high Reynolds numbers, the filter width will be very large in comparison to the Kolmogorov length scale and it can be expected that $k'_{u@p}$ becomes dominant in comparison to $\overline{\langle u_{f@p,i} \rangle u'_{f@p,i}}$ in this case.

Furthermore, figure 5.15 shows that at small Reynolds numbers, $k'_{u@p}$ shows little Stokes number dependence but with increasing Reynolds number, Stokes number dependence evolves. For typical LES it can be expected that $k'_{u@p}$ becomes strongly dependent on Stokes number. Shotorban & Mashayek (2006) and Simonin *et al.* (1993) assume for their models that $k'_{u@p}$ is independent of Stokes number. The present results indicate that this assumption is questionable.

Figure 5.15 shows further that particles with $St \approx 1$ tend to cluster in regions with lower SGS kinetic energy than the ones with very small or very large Stokes numbers. They even cluster in regions with sub-average SGS kinetic energy. At the same time it is well known that preferential concentration is maximised around $St = 1$. This means that one might model $k'_{u@p}$ in dependence of Stokes number. However, to date there is no physical explanation for such a model. Furthermore, this hypothesis would need to be verified by simulations at even higher Reynolds number. Only if these two issues are solved, then a reliable model can be constructed.

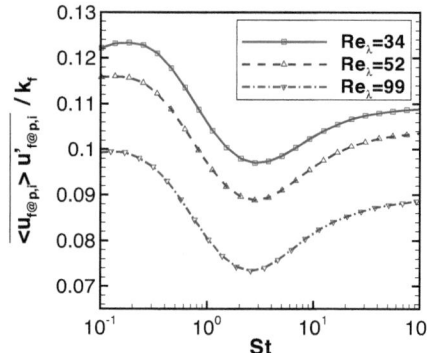
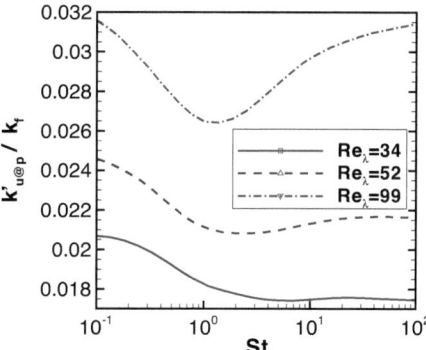

Figure 5.14: A priori analysis: Covariance of filtered velocity and fluctuations seen by the particles.

Figure 5.15: A priori analysis: Kinetic energy of the fluctuations seen by the particles.

Underprediction of kinetic energy of the particles in LES

For most applications the kinetic energy seen by the particles is not as important as the kinetic energy of the particles themselves. In the following it is shown that this quantity is underpredicted in LES as expected.

Figure 5.16 shows the kinetic energy of the particles in DNS and LES

$$k_p = \frac{1}{2}\overline{u_{p,i}^2} \quad \text{and} \quad [k_p] = \frac{1}{2}\overline{[u_{p,i}]^2}. \tag{5.16}$$

For all Reynolds and Stokes numbers, $[k_p]$ is smaller than k_p. For $St \to \infty$, $[k_p]$ converges towards k_p; the effect of the small scale turbulence onto k_p becomes negligible.

An analytical estimate for $k_p/k_{u@p}$ can be obtained by assuming an exponential form of the autocorrelation function of fluid velocity seen by particles (see Fede & Simonin, 2006). This estimate reads

$$\frac{k_p}{k_{u@p}} = \frac{1}{1+St_\eta} \quad \text{with } St_\eta = \frac{1}{t_{u@p}}\left(\overline{\frac{c_D Re_p}{24\tau_p}}\right)^{-1}. \tag{5.17}$$

$t_{u@p}$ is the Lagrangian integral time scale of the fluid velocity seen by the particle (cf. also section 5.5.2). If a biexponential form of the autocorrelation function of fluid velocity seen by particles is assumed, then the following estimate can be obtained (see Fede & Simonin, 2006):

$$\frac{k_p}{k_{u@p}} = \frac{2\,St_\eta z^2}{2\,St_\eta + 2\,St_\eta^2 + z^2} \tag{5.18}$$

z is the ratio between Taylor and Lagrangian time scale. These two estimates are depicted in figure 5.17 together with results from the numerical simulations. Apparently, DNS and LES produce at all Reynolds numbers results close to these two estimates. At the highest Reynolds number, however, slight deviations can be found for large values of St_η in LES and DNS.

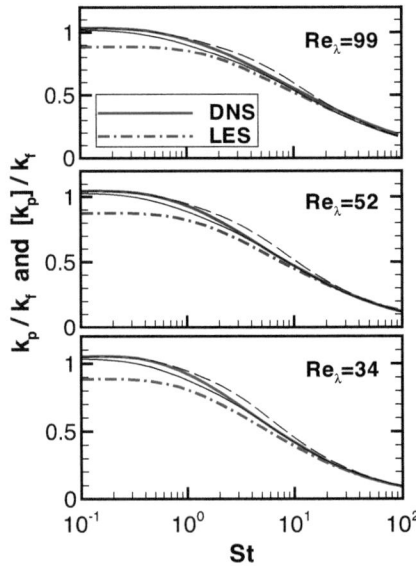

Figure 5.16: A posteriori analysis: Kinetic energy of the particles. Bold continuous lines: DNS, bold dash-dotted lines: LES, thin continuous lines: estimate from equation (5.17), thin dashed lines: estimate from equation (5.18).

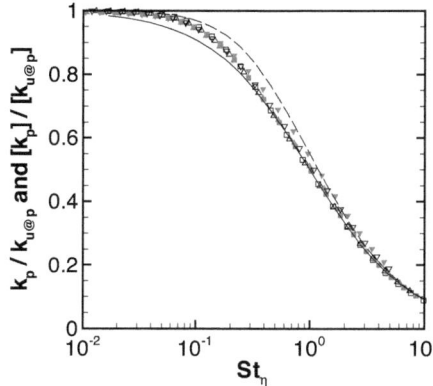

Figure 5.17: A posteriori analysis: Kinetic energy of the particles with respect to the kinetic energy of the fluid seen by the particles. Continuous line: estimate from equation (5.17), dashed line: estimate from equation (5.18). Filled symbols: LES, hollow symbols: DNS. Square symbols: $Re_\lambda = 34$, delta symbols: $Re_\lambda = 52$, inverse delta symbols: $Re_\lambda = 99$.

5.5.2 Integral time scale

This section contains an analysis of Lagrangian integral time scales of fluid velocity seen by the particles and of particle velocity. The former is a crucial parameter for Langevin-based particle SGS models (see Shotorban & Mashayek, 2006; Simonin et al., 1993; Fede et al., 2006) whereas the latter determines particle dispersion (see Taylor, 1922). Thus, both quantities are important for modelling issues.

Scaling law for the integral time scale of the fluid seen by the particles

Figure 5.18 shows the integral time scales of the fluid velocity seen by the particles computed from unfiltered and filtered DNS data

$$t_{u@p} = \int_0^\infty \frac{\overline{u_{f@p,i}(t) u_{f@p,i}(t+s)}}{\overline{u_{f@p,i}^2(t)}} \, ds \tag{5.19a}$$

$$\langle t_{u@p} \rangle = \int_0^\infty \frac{\overline{\langle u_{f@p,i} \rangle (t) \langle u_{f@p,i} \rangle (t+s)}}{\overline{\langle u_{f@p,i} \rangle (t)^2}} \, ds. \tag{5.19b}$$

For $St \to \infty$, this time scale approaches the Eulerian integral time scale of the flow and for $St = 0$ this time scale is equivalent to the Lagrangian time scale of particles which follow the fluid exactly. The theory of Sawford (1991) gives an estimate for the latter particles:

$$t_{u@p} = t_p(St = 0) = \tau_k \frac{2 Re_\lambda}{C_0 \sqrt{15}} (1 + 7.5 \, C_0^2 \, Re_\lambda^{-1.64}), \tag{5.20}$$

where C_0 is the Kolmogorov constant at infinite Reynolds number. Rodean (1991) deduced a value of $C_0 = 5.7$ from theoretical considerations but experiments (see Mordant et al.,

2001; Reynolds, 2003; Ouellette et al., 2006; Lien & D'Asaro, 2002; Sawford, 1991; Pope, 2000) show that C_0 can range from 4 to 7 . Reynolds number dependence can be included following the recommendations of Fox & Yeung (2003):

$$C_0 = 6.5 \left(1 + \frac{8.1817}{Re_\lambda}\left(1 + \frac{110}{Re_\lambda}\right)\right)^{-1}. \tag{5.21}$$

This formula was obtained by fitting data from DNS of isotropic turbulence. The estimates from this equation are presented in table 5.3 for the present Reynolds numbers.

Table 5.3: Kolmogorov constant C_0 following Fox & Yeung (2003) (equation (5.21)).

Re_λ	34	52	99	265
C_0	3.22	4.36	5.53	6.23

In figure 5.18 the estimate from equation (5.20) is plotted for $C_0 \in [4, 5.7]$. It can be seen that in the present study DNS produces results within these limits at all Reynolds numbers.

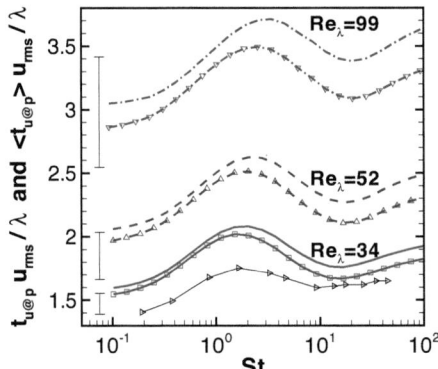

Figure 5.18: A priori analysis: Integral time scale of unfiltered and filtered fluid velocity seen by the particles. Bold continuous line: $Re_\lambda = 34$, dashed line: $Re_\lambda = 52$, dash-dotted line: $Re_\lambda = 99$. Thin continuous line: Data of Fede & Simonin (2006) (unfiltered DNS at $Re_\lambda = 34.1$). Lines with symbols: DNS, lines without symbols: filtered DNS. Error bars: Estimates for $t_{u@p}(St = 0)$ according to equation (5.20).

Filtering leads to a smoother field and therefore filtering increases the integral time scale $t_{u@p}$ at all Reynolds and Stokes numbers. The effect of filtering increases with Reynolds number (the reader is reminded that the filter was chosen such that for all Reynolds numbers the percentage of resolved kinetic energy is approximately equal). This means that the small scales have a significant effect on $t_{u@p}$ at all Stokes numbers.

In accordance with the findings of Fede & Simonin (2006), $t_{u@p}$ shows a local maximum and a local minimum due to clustering. Section 5.5.1 showed already that particles with $1 < St \lesssim 10$ cluster in regions of low kinetic energy. Figure 5.18 shows that these particles additionally see longer integral time scales in the surrounding flow. This result is in accordance with the results of Chen, Goto & Vassilicos (2006) and Goto & Vassilicos

(2006). These authors found that particles tend to cluster in zero acceleration points. A particle-SGS model that reconstructs the seen fluid velocity would have to respect this, accordingly.

LES produces qualitatively the same result. Figure 5.19 shows the integral time scale of the fluid velocity seen by the particles in LES,

$$[t_{u@p}] = \int_0^\infty \frac{\overline{[u_{f@p,i}(t)][u_{f@p,i}(t+s)]}}{\overline{[u_{f@p,i}]^2(t)}}\, ds. \tag{5.22}$$

Additionally the corresponding quantities from DNS and filtered DNS are shown. Apparently $|[t_{u@p}] - t_{u@p}|$ is greater than $|\langle t_{u@p}\rangle - t_{u@p}|$, i.e., the integral time scale computed from LES data shows greater error than the integral time scale computed from filtered DNS data.

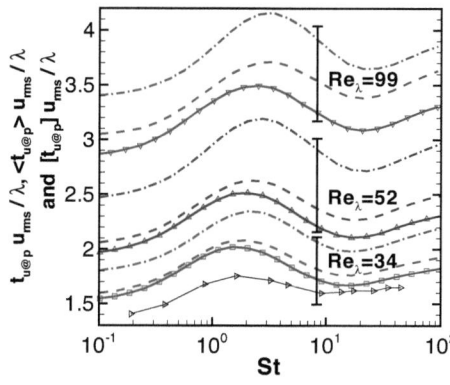

Figure 5.19: Integral time scale of the fluid velocity seen by the particles in DNS, filtered DNS and LES. Continuous lines: DNS, dashed lines: filtered DNS, dash-dotted lines: LES. Thin continuous line with symbols (lowermost line): data of Fede & Simonin (2006), unfiltered DNS at $Re_\lambda = 34.1$.

This might be due to two effects, namely either because of different particle paths in filtered DNS and LES or because of approximation errors of the fluid-LES model, i.e., differences in the statistics of filtered DNS and LES velocity. In order to separate these effects, one DNS at $Re_\lambda = 99$ was conducted where particles with $St = 0.1$ were traced along the particle path computed from the filtered DNS result. This means that here the second effect, an approximation error of the fluid-LES model, was excluded.

Figure 5.20 shows the autocorrelation of the fluid velocity seen by the particles

- from the original DNS,

- from the filtered field recorded along DNS particle paths,

- from the filtered field recorded along particle paths which were computed from the filtered field

- and from LES.

Evidently the correlation function from filtered particle paths is very much closer to the DNS result than to the LES result. This means that the different particle paths do not affect the integral time scale as strongly as approximation errors from LES. With other words, the eddy life time in LES is larger than the eddy life time of the filtered fluid velocity, both computed along the particle paths which the large scales prescribe.

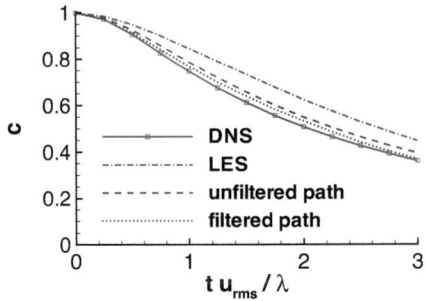

Figure 5.20: Autocorrelation of fluid velocity seen by particles at $Re_\lambda = 99$, $St = 0.1$. Continuous line: DNS, dash-dotted line: LES, dashed line: filtered DNS velocity, recorded along path from unfiltered field, dotted line: filtered DNS velocity, recorded along path from filtered field.

Concluding, it is clear that a particle-LES model must decrease the integral time scale of the fluid velocity seen by the particles. On the other hand, with the present fluid-LES model the DNS result can only be obtained if the particle-LES model modifies even the resolved scales. It is questionable whether the better alternative might be to improve the fluid-LES model.

For the kinetic energy seen by the particles, two scaling laws were formulated (law 1 and 2). They concern the scaling of the kinetic energy seen by the particles in dependence of the particle relaxation time. Concerning the integral time scale, a similar rescaling can be conducted. Figure 5.21 shows the integral time scale of the fluid velocity seen by the particles in DNS and filtered DNS, rescaled with respect to the time scale at $St = 1$. The corresponding LES result is shown in figure 5.22. Stokes number is again based on the smallest resolved time scale. In constrast to the kinetic energy, figures 5.9 and 5.10, the integral time scale does not show a linear regime around $St = 1$ but the tangent of all lines is identical. In a first-order approach, only the tangent is of interest. This allows to formulate the following scaling law.

Scaling Law 3. *Scaling of integral time seen by particles at small relaxation time. Define the Stokes number St_S as in scaling law 1, i.e.,*

$$\text{in DNS set } St_S = St = \frac{\tau_p}{\tau_K} = \frac{\tau_p\sqrt{\epsilon}}{\sqrt{\nu}}, \quad \text{in LES set } St_S = \frac{\tau_p\sqrt{[\epsilon]}}{\sqrt{\nu+\nu_t}}. \tag{5.23}$$

Then, $t_{u@p}/t_{u@p}(St_S = 1)$, $\langle t_{u@p}\rangle / \langle t_{u@p}(St_S = 1)\rangle$ and $[t_{u@p}]/[t_{u@p}(St_S = 1)]$ scale in first-

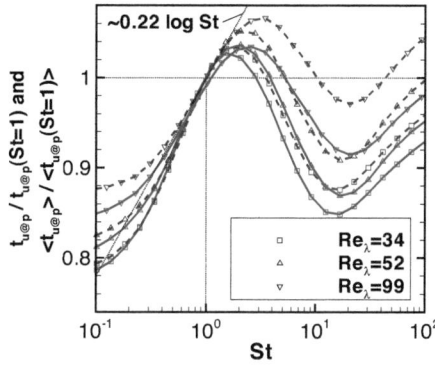

Figure 5.21: A priori analysis: Integral time scale of the fluid seen by the particles computed from DNS, scaled by $t_{u@p}(St = 1)$ and $\langle t_{u@p}(St = 1)\rangle$, respectively. Continuous lines: DNS, dashed lines: filtered DNS.

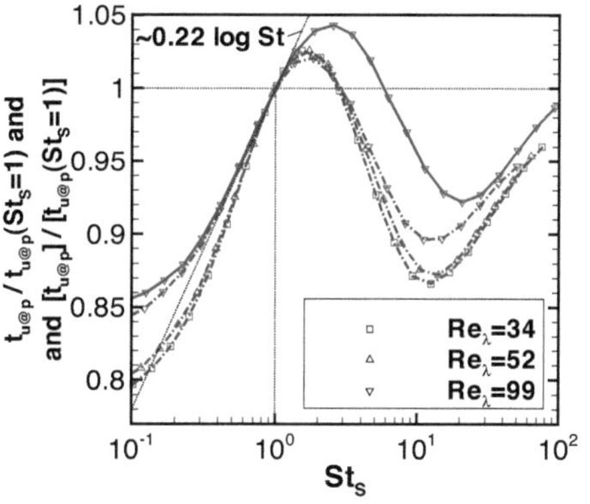

Figure 5.22: A posteriori analysis: Integral time scale of the fluid seen by the particles, scaled by $t_{u@p}(St = 1)$ and $[t_{u@p}(St = 1)]$, respectively. Scaling on x-axis is based on smallest resolved time scale. Continuous line: DNS at $Re_\lambda = 99$, dash-dotted lines: LES.

order with $\log St_S$ around $St_S = 1$. The scaling factor is approximately 0.22,

$$\left.\begin{array}{l} \dfrac{t_{u@p}}{t_{u@p}(St_S=1)} \sim 0.22 \log \dfrac{\tau_p \sqrt{\epsilon}}{\sqrt{\nu}}, \\[2mm] \dfrac{\langle t_{u@p}\rangle}{\langle t_{u@p}(St_S=1)\rangle} \sim 0.22 \log \dfrac{\tau_p \sqrt{\epsilon}}{\sqrt{\nu}}, \\[2mm] \dfrac{[t_{u@p}]}{[t_{u@p}(St_S=1)]} \sim 0.22 \log \dfrac{\tau_p \sqrt{[\epsilon]}}{\sqrt{\nu+\nu_t}} \end{array}\right\} \text{ around } St_S \approx 1. \qquad (5.24)$$

Scaling law 3 corresponds to scaling law 1. Both concern particle dynamics around $St_S = 1$. Law 2 concerns scaling of kinetic energy around $St_L = 1$. In order to analyze the scaling of the integral time around $St_L = 1$, one plots the integral time against St_L, figures 5.23 and 5.24. From this data no scaling law around St_L can be deduced, thus there is no one to one correspondence between law 2 and the respective formulation for the integral time.

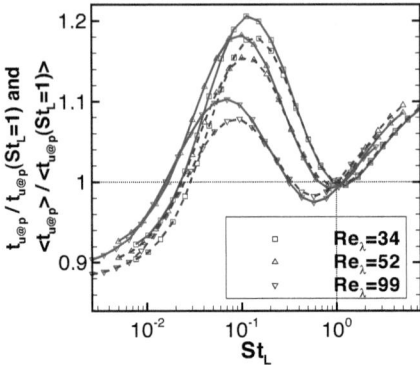 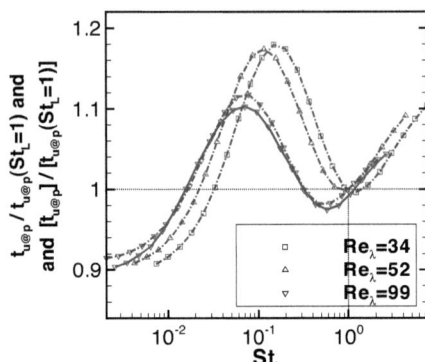

Figure 5.23: A priori analysis: Integral time scale of the fluid seen by the particles computed from DNS, scaled by $t_{u@p}(St_L = 1)$ and $\langle t_{u@p}(St_L = 1)\rangle$, respectively. Scaling on x-axis is based on large eddy decay time. Continuous lines: DNS, dashed lines: filtered DNS.

Figure 5.24: A posteriori analysis: Integral time scale of the fluid seen by the particles, scaled by $t_{u@p}(St_L = 1)$ and $[t_{u@p}(St_L = 1)]$, respectively. Scaling on x-axis is based on large eddy decay time. Continuous line: DNS at $Re_\lambda = 99$, dash-dotted lines: LES.

Stokes number dependence of SGS integral time scale seen by the particles

On page 64, the SGS kinetic energy seen by the particles was analysed because this quantity is a parameter for the models of Shotorban & Mashayek (2006) and Simonin et al. (1993) (cf. chapter 6). The second parameter for these models is the SGS integral time seen by the particles. Similar to the kinetic energy, both models assume that the SGS integral time scale is independent of Stokes number. In the following it is shown that this assumption is questionable.

The integral time scale of the fluctuations seen by the particles is defined by

$$t'_{u@p} = \int_0^\infty \frac{\overline{u'_{f@p,i}(t)u'_{f@p,i}(t+s)}}{\overline{u'_{f@p,i}(t)u'_{f@p,i}(t)}}\,ds. \tag{5.25}$$

The particle LES models by Shotorban & Mashayek (2006) and by Simonin et al. (1993) need an estimate for $t'_{u@p}$. Currently, information about the Stokes number dependence of $t'_{u@p}$ is unavailable. Therefore, for both models the authors assume that $t'_{u@p}$ is independent of Stokes number. Figure 5.25 shows that in the configurations analysed the validity of this assumption is questionable. It seems that the effect of St on $t'_{u@p}$ is stronger at higher Reynolds

5 A numerical study on requirements for a particle-LES model

numbers, which introduces an additional difficulty into modelling.

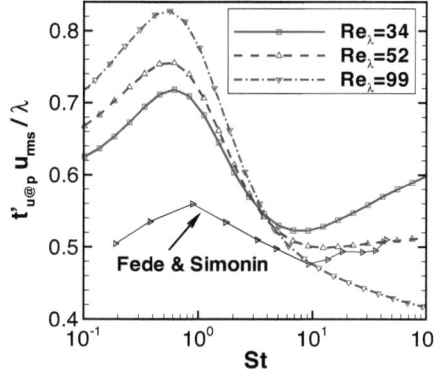

Figure 5.25: A priori analysis: Integral time scale of the small scale fluid velocity seen by the particles. Bold continuous line: $Re_\lambda = 34$, dashed line: $Re_\lambda = 52$, dash-dotted line: $Re_\lambda = 99$. Thin continuous line: Data of Fede & Simonin (2006) (spectral filter, $Re_\lambda = 34.1$).

Overprediction of the integral time scale of the particle velocity in LES

In the following, SGS effect on the integral time scale of the particle velocity are analysed. This quantity is defined by

$$t_p = \int_0^\infty \frac{\overline{v_i(t)v_i(t+s)}}{\overline{v_i^2(t)}} \, ds \quad \text{and} \quad [t_p] = \int_0^\infty \frac{\overline{[v_i](t)[v_i](t+s)}}{\overline{[v_i]^2(t)}} \, ds \qquad (5.26)$$

for DNS and LES, respectively. Both quantities are plotted in figure 5.26. For all Reynolds and Stokes numbers, the integral time scale computed from LES is higher than the integral time scale from DNS, in accordance with the results of Yang et al. (2008).

As expected, LES and DNS results collapse for high Stokes numbers. For LES this means that, concerning t_p, the effect of the SGS turbulence can be safely neglected for large Stokes numbers. For low Stokes numbers, LES shows larger integral time scales than DNS.

5.5.3 Particle dispersion

For applications, often SGS effect on the rate of dispersion are very relevant. This is analysed in the following.

As mentioned above, long time particle dispersion D can be computed from the product of kinetic energy and integral time scale,

$$D = \lim_{s \to \infty} \frac{d\|\mathbf{x}_p(t+s) - \mathbf{x}_p(t)\|^2}{ds} = 4k_p t_p. \qquad (5.27)$$

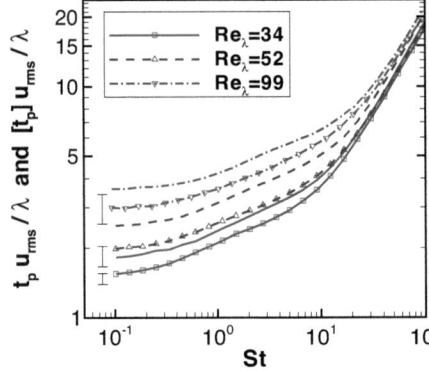

Figure 5.26: A posteriori analysis: Integral time scale of particle velocity. Lines with symbols: DNS, lines without symbols: LES. Errorbars: Estimates for $t_p(St = 0)$ according to equation (5.20).

D is plotted in figure 5.27. At the smallest Reynolds number, small scale effects on the kinetic energy and small scale effects on integral time scale cancel out each other. This is in accordance with the findings of Fede & Simonin (2006).

On the other hand, for the higher Reynolds numbers ($Re_\lambda = 52$ and $Re_\lambda = 99$), small scale effects on the integral time scale are stronger than on the kinetic energy, resulting in a higher rate of dispersion in LES than in DNS. This suggests that for very high Reynolds numbers one cannot assume that LES predicts dispersion correctly.

As discussed in section 5.5.2, this might also be an effect of the fluid-LES model. In particular, at low Reynolds number or highly resolved LES other authors also reported that dispersion is predicted correctly in LES (see e.g. Armenio et al., 1999; Fede & Simonin, 2006; Yang et al., 2008). Therefore the present finding extends their findings in the sense that, concerning rate of dispersion, SGS effects in highly resolved LES and in coarse LES can differ qualitatively.

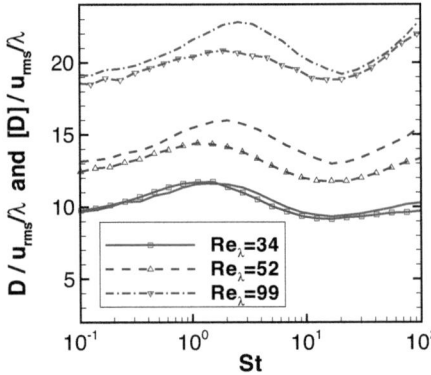

Figure 5.27: A posteriori analysis: Turbulent particle dispersion. Lines with symbols: DNS, lines without symbols: LES.

5.5.4 Preferential concentration

In section 2.2.3 the effect of preferential concentration was explained. In that section, also two measures for preferential concentration, namely accumulation Σ and fractal dimension d_{pc} were presented. In the present section the effect of the unresolved scales on preferential concentration is analysed.

In order to analyse preferential concentration, the number of particles must be large enough to resolve the scale on which clustering occurs. In isotropic turbulence, these scales are known to be in the range of $2 - 6\eta_K$ (cf. e.g. Hogan & Cuzzi, 2001). Therefore in the present work Σ was computed by using boxes of dimensions $(3\eta_K)^3$. Here, 800,000 particles per fraction were traced for $Re_\lambda = 34$ and 5 Mio. particles per fraction for $Re_\lambda = 52$. Then, in a box of dimensions $(3\eta_K)^3$ one will find on average 3.4 particles.

In these simulations, the particles' Stokes numbers range from 0.1 to 10. In each simulation, 10 time samples were taken. The time lag between the samples is approximately 10 Kolmogorov time scales.

Meyer & Jenny (2004) pointed out that non-conservative interpolation of the fluid velocity on the particle position can lead to artificial particle clustering. Therefore all simulations for preferential concentration were run with two different interpolation methods, namely a standard Lagrangian (non-conservative) fourth-order interpolation and a conservative second-order interpolation scheme. The results do not differ substantially, therefore in the following only the results from fourth-order interpolation are shown.

In figure 5.28 the accumulation Σ and the fractal dimension d_{pc} is depicted for both Reynolds numbers. Evidently DNS and LES results differ somewhat but nevertheless LES gives qualitatively the same result, namely that preferential concentration is strongest around $St = 1$. This is in accordance with the findings of Wang & Squires (1996) who also found that LES predicts preferential concentration quite well.

At this point, the results concerning preferential concentration are not very interesting. However, it will be shown later that models which enhance single particle statistics may destroy preferential concentration, cf. section 6.4.4.

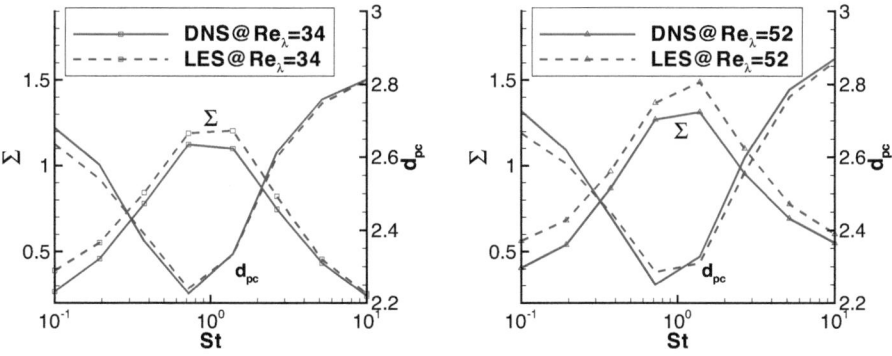

Figure 5.28: A posteriori analysis of preferential concentration: Continuous lines: DNS, dashed lines: LES. Lines with symbols: Σ (left axis), lines without symbols: d_{pc} (right axis).

5.6 Conclusions of chapter 5

Chapter 5 contains an investigation of the effect of small scale turbulence on suspended particles. These effects need to be emulated by a particle-LES model. The investigation was conducted by a priori and a posteriori analysis of homogeneous isotropic turbulence at Reynolds numbers $Re_\lambda = 34$, 52 and 99. Stokes numbers based on the Kolmogorov time scale range from 0.1 to 100.

The analysis focuses on the criteria stated in chapter 4, namely kinetic energy of fluid seen by the particles, kinetic energy of the particles, rate of dispersion and preferential concentration. Additionally, the integral time scales are analysed.

The results presented here show that the kinetic energy of suspended particles is smaller in LES than in DNS. On the other hand, the integral time scale of particle velocity is higher in LES than in DNS. In consequence, the rate of dispersion, being the product of kinetic energy and integral time scale, shows smaller differences between LES and DNS.

At small Reynolds number ($Re_\lambda = 34$), the rate of dispersion is predicted correctly by LES. This is in accordance with the findings of other authors (see e.g. Armenio et al., 1999; Fede & Simonin, 2006; Yang et al., 2008). On the other hand, the present results show that at high Reynolds numbers ($Re_\lambda = 52$ and 99), the overprediction of the integral time scale in LES is stronger than the underprediction of the kinetic energy, resulting in a higher rate of dispersion in LES than in DNS. This suggests that for very high Reynolds numbers one cannot assume that LES predicts dispersion correctly.

This shows that in general modelling is necessary. The most promising models which were published to date are based on the reconstruction of the fluid velocity seen by the particles $\mathbf{u}_{f@p}$, cf. chapter 6. In particular, some models take the kinetic energy of the unresolved scales $k'_{u@p}$ as input parameter (e.g. the models of Shotorban & Mashayek (2006) and Simonin et al. (1993)). The present study shows that clustering has a significant effect on this quantity. This complicates modelling especially if clustering is determined by non-resolved scales (cf. Fede & Simonin, 2006).

Furthermore, the study shows that particles with intermediate relaxation times tend to cluster in regions with low turbulent kinetic energy. The relaxation time where this effect is maximised is significantly larger than the Kolmogorov time scale. Results indicate furthermore that locations for clustering are determined by large scales. The mechanisms leading to clustering are determined by large and small scales. The strong influence of the large scales on clustering is a promising result for LES because even if the dissipative scales are unresolved then still the effect of clustering on velocity statistics might be predicted almost correctly.

As a by-product, the present study showed that the kinetic energy seen by the particles as a function of Stokes number can be divided in two regimes such that the first regime is governed by the Kolmogorov time scale and the second regime is governed by the large eddy decay time. In each regime, the Stokes number dependence was described by a scaling law. The law is valid for DNS, filtered DNS and LES. Concerning integral time scale seen by the particles, only one regime could be identified, namely the regime governed by the Kolmogorov time.

Furthermore, in the present configurations the effect of small scale turbulence on preferential concentration was found to be negligible. This means on the one hand that a particle-LES model does not need to reconstruct mechanisms leading to preferential concentration. On

5 A numerical study on requirements for a particle-LES model 77

the other hand, a particle-LES model must preserve preferential concentration. In the following chapter it will be shown that the models proposed by Shotorban & Mashayek (2006) and Simonin *et al.* (1993) do not fulfil this requirement.

6 Presentation and statistical assessment of existing particle-LES models

> *In theory, there is no difference between theory and practice. But in practice, there is.*
>
> *Chuck Reid*

The works of Armenio *et al.* (1999); Yamamoto *et al.* (2001); Fede & Simonin (2006) and the results form the previous chapter show that in general SGS effects on particles cannot be neglected. Thus, the need for a corresponding model arises.

In this chapter, three such models are analysed in detail with respect to the criteria listed in chapter 4. The analysis focuses on the Approximate Deconvolution Method (ADM) as proposed by Kuerten (2006*b*) and two Langevin-based models proposed by Shotorban & Mashayek (2006) and Simonin *et al.* (1993).

In the following, first a notation is introduced and an overview on available particle-LES models is given. Then, results from analytical and numerical assessment of the models are presented.

6.1 A word on notation in this chapter

In the previous chapters, the notation '@p' stood only for the fluid velocity at the particle position $\mathbf{u}_{f@p}$. In the present chapter, the notation '@p' is adopted for arbitrary functions $f(\mathbf{x}, t)$, i.e.,

$$f_{@p}(t) = f\left(\mathbf{x}_p(t), t\right). \tag{6.1}$$

For example $\mathbf{u}_f$ refers to the space- and time-dependent solution of the Navier–Stokes equations whereas $\mathbf{u}_{f@p}$ refers to the time-dependent fluid velocity seen by the particle. Correspondingly, $\mathcal{G}\mathbf{u}_f$ refers to the space- and time-dependent solution of the filtered Navier–Stokes equations whereas $(\mathcal{G}\mathbf{u}_f)_{@p}$ refers to the time-dependent filtered fluid velocity seen by the particle.

6.2 Particle-LES models

It is clear that there cannot be one model which is the best choice for all applications. For example, in general a model with high accuracy will be expensive in terms of CPU time. CPU time is limited and therefore one must decide between accuracy of the particle-LES model and grid refinement. The present work focuses on assessment of particle-LES models with respect to their accuracy, neglecting the computational requirements. In this section, several particle-LES models are presented.

6.2.1 Overview on particle-LES models

One of the first particle-LES models for inert particles was proposed by Simonin, Deutsch & Minier (1993). Their model is based on a model for inertia free particles by Haworth & Pope (1986). With this model, Simonin *et al.* generated reference data in order to construct a Eulerian model for the correlations between particle and fluid velocity. In 1996, Wang & Squires (1996) proposed another particle-LES model. This model is based on a transport equation for the unresolved kinetic energy and the authors used this model in order to predict preferential concentration in turbulent channel flow. In recent years, development of particle-LES models has increased rapidly. Within the last 3 years, at least seven new models (see Shotorban & Mashayek, 2006; Kuerten, 2006b; Amiri, Hannani & Mashayek, 2006; Gobert, Motzet & Manhart, 2007; Shotorban, Zhang & Mashayek, 2007; Bini & Jones, 2007, 2008) were proposed.

Most of the models mentioned are stochastic models. These are often obtained by extending models developed for Reynolds Averaged Navier–Stokes (RANS) simulations. Among these stochastic models is the large class of Langevin-based models (see Haworth & Pope, 1986; Pope, 2000). For these models, stochastic differential equations need to be solved.

An intrinsic problem of stochastic models is that the modelled quantity (such as particle position or particle velocity) is not differentiable. This means that statistics of time derivatives of the modelled quantity are not available. Stochastic models for higher order time derivatives (i.e., particle acceleration or second derivative of particle velocity) circumvent this problem to a certain extent but introduce other unphysical effects, such as incorrect particle velocity spectra (see Sawford, 1991).

Problems of differentiability can be circumvented by deterministic models. A very promising deterministic model is the approximate deconvolution method (ADM) for particle-laden flows (see Kuerten, 2006b; Shotorban *et al.*, 2007; Shotorban & Mashayek, 2005). Here, the barely resolved scales are improved significantly. However, scales which are smaller than the LES grid cannot be reconstructed with this method (see Kuerten, 2006b).

6.2.2 Analysed particle-LES models

Section 5.5.1 showed already that, except for the sharp spectral filter, the filter transfer function $\mathcal{FT}(G)$ decays continuously from 1 at wavenumber 0 to zero at the highest resolved wavenumber κ_c. Thus, 'reconstruction of the unresolved scales' actually has two meanings: (a) reconstruction of scales which are resolvable on the LES grid but damped due to filtering and (b) reconstruction of the effect of scales smaller than the LES grid. ADM addresses the first issue and stochastic models rather focus on the second issue.

Approximate Deconvolution Method (ADM)

ADM is well established for incompressible single phase flows (see Stolz & Adams, 1999; Schlatter, 2004; Stolz, Adams & Kleiser, 2001b). Kuerten (2006a,b), Shotorban & Mashayek

(2005) and Shotorban et al. (2007) analysed the capabilities of ADM for particle-laden flow. With ADM, the fluid velocity seen by the particle $\mathbf{u}_{f@p}^{ADM}$ is computed from

$$\mathbf{u}_{f@p}^{ADM} = \left(\mathbf{u}_f^{ADM}\right)_{@p} = \sum_{n=0}^{N} \left((\mathcal{I} - \mathcal{G})^n \mathcal{G} \mathbf{u}_f\right)_{@p}. \tag{6.2}$$

Here, $\mathcal{I}$ stands for identity. N is the number of deconvolution steps. Equation (6.2) is solved once per time step and the particle velocity is computed from

$$\frac{\mathrm{d}\mathbf{u}_p^{ADM}}{\mathrm{d}t} = \frac{c_D Re_p}{24\tau_p} \left(\mathbf{u}_{f@p}^{ADM} - \mathbf{u}_p^{ADM}\right). \tag{6.3}$$

The operator $\mathcal{H} = \mathcal{I} - \mathcal{G}$ can be interpreted as extractor of subgrid scales. With this operator, $\mathbf{u}_f^{ADM}$ can be written as

$$\mathbf{u}_f^{ADM} = \sum_{n=0}^{N} \mathcal{H}^n \mathcal{G} \mathbf{u}_f = \sum_{n=0}^{N} \mathcal{H}^n (\mathcal{I} - \mathcal{H}) \mathbf{u}_f = \left(\mathcal{I} - \mathcal{H}^{N+1}\right) \mathbf{u}_f = \mathcal{H}^{ADM} \mathbf{u}_f. \tag{6.4}$$

For $N \to \infty$ the transfer function of $\mathcal{H}^{N+1}$ equals zero for the resolvable scales ($\|\mathbf{k}\| < \kappa_c$) and one for the unresolvable scales ($\|\mathbf{k}\| > \kappa_c$). This shows that for large N, the effect of ADM can be interpreted as improving the LES filter towards a sharp spectral filter.

Langevin-based model proposed by Shotorban & Mashayek.

ADM cannot reconstruct scales smaller than the LES grid. In order to circumvent this, Shotorban & Mashayek (2006) propose a stochastic model based on a Langevin equation for the fluid velocity seen by a particle. This model is based on works for single phase or reacting flows by Pope (1983), Heinz (2003) and Gicquel et al. (2002). These authors proposed and analysed stochastic models for the Lagrangian fluid velocity for inertia free particles. These models are called generalised Langevin models.

Shotorban & Mashayek adopted this type of model for inert particles. They propose to compute the fluid velocity seen by the particles $\mathbf{u}_{f@p}^{Sho}$ from the stochastic differential equation (Langevin equation)

$$\mathrm{d}u_{f@p,i}^{Sho} = \left(\mathcal{G}\left(\frac{\partial u_{f,i}}{\partial t} + u_{f,j}\frac{\partial u_{f,i}}{\partial x_j}\right)\right)_{@p} \mathrm{d}t - \frac{u_{f@p,i}^{Sho} - (\mathcal{G}u_{f,i})_{@p}}{T_L} \mathrm{d}t + \sqrt{C_0 \epsilon}\ \mathrm{d}W_i. \tag{6.5}$$

The reader is reminded that '@p' denotes 'at the particle position', cf. equation (6.1). The first term on the right hand side of equation (6.5) is the filtered material derivative of the fluid velocity and can be computed from the right hand side of the filtered Navier–Stokes equation. The second term is a drift term for the random variable $\mathbf{u}_{f@p}^{Sho}$, leading to a relaxation of $\mathbf{u}_{f@p}^{Sho}$ against $(\mathcal{G}\mathbf{u}_f)_{@p}$. The last term is a diffusion term for $\mathbf{u}_{f@p}^{Sho}$. $\mathbf{W}$ denotes a Wiener process and ϵ is the (modelled) dispersion of subgrid scale kinetic energy.

Two parameters need to be specified, namely the time scale T_L and the Kolmogorov con-

6 Presentation and assessment of existing particle-LES models

stant C_0 (cf. section 5.5.2). For single phase flows, the choice

$$T_L = \frac{k_{sgs}}{\left(\frac{1}{2} + \frac{3}{4}C_0\right)\epsilon} \tag{6.6}$$

with the subgrid kinetic energy k_{sgs} guarantees that the rate of dissipation predicted by the model is correct in decaying isotropic turbulence (see Pope, 2000). C_0 can be set such that experimental data from a thermal wake are fitted well (see Pope, 2000). This gives $C_0 = 2.1$, in good accordance with the experimental findings of Walpot et al. (2007) in turbulent channel flow. In isotropic turbulence, the relation proposed by Fox & Yeung (2003) (cf. equation (5.21)) can be used.

Shotorban & Mashayek propose to compute the particle velocity from

$$\mathrm{d}\mathbf{u}_p^{Sho} = \frac{c_D Re_p}{24\tau_p}\left(\mathbf{u}_{f@p}^{Sho} - \mathbf{u}_p^{Sho}\right)\,\mathrm{d}t. \tag{6.7}$$

The model is closed by additional estimates for k_{sgs} and ϵ.

Berrouk et al. (2007) propose a similar model where C_0 is computed from flow statistics. In the present analysis, C_0 is not restricted to any value. Therefore the present analysis holds for the model proposed by Berrouk et al. as well.

Langevin-based model proposed by Simonin et al.

Simonin et al. (1993) also propose to model the fluid velocity seen by the particles by a stochastic process. Fede et al. (2006) presented in detail how to deduct Simonin et al.'s model for particle-laden flow starting from the Navier–Stokes equations. This results in a different Langevin equation than the equation proposed by Shotorban & Mashayek.

In contrast to Shotorban & Mashayek, Simonin et al. propose to transport the resolved scales by particle velocity (and not by fluid velocity). The model can be formulated via a Langevin equation for the unresolved scales

$$\mathrm{d}u_{f@p,i}^{Sim'} = \left(-u_{f@p,j}^{Sim'}\left(\frac{\partial \mathcal{G}u_{f,i}}{\partial x_j}\right)_{@p} + \left(\frac{\partial \tau_{i,j}}{\partial x_j}\right)_{@p} + \Gamma_{ij}u_{f@p,j}^{Sim'}\right)\,\mathrm{d}t + \sqrt{C_0\epsilon}\,\mathrm{d}W_i. \tag{6.8}$$

$\tau_{ij} = \mathcal{G}\left(u_i u_j\right) - \mathcal{G}u_i \mathcal{G}u_j$ is the SGS stress tensor. The matrix Γ is comparable to $-1/T_L$, T_L being the relaxation time scale of the model of Shotorban & Mashayek. The model constant C_0 is equivalent to C_0 of Shotorban & Mashayek's model.

The fluid velocity seen by the particles is computed from $\mathbf{u}_{f@p}^{Sim} = \left(\mathcal{G}\mathbf{u}_f\right)_{@p} + \mathbf{u}_{f@p}^{Sim'}$. This is equivalent to solving

$$\mathrm{d}\mathbf{u}_{f@p}^{Sim} = \mathrm{d}\left(\mathcal{G}\mathbf{u}_f\right)_{@p} + \mathrm{d}\mathbf{u}_{f@p}^{Sim'}. \tag{6.9}$$

Then, the particle velocity is computed from

$$\mathrm{d}\mathbf{u}_p^{Sim} = \frac{c_D Re_p}{24\tau_p}\left(\mathbf{u}_{f@p}^{Sim} - \mathbf{u}_p^{Sim}\right)\,\mathrm{d}t. \tag{6.10}$$

In the context of Reynolds averaged Navier–Stokes simulations (RANS), similar models were analysed and successfully used in many works (see Oesterlé & Zaichik, 2004; Beishuizen, Naud & Roekaerts, 2007; Minier, Peirano & Chibbaro, 2004; Sawford, 2001). There, a common choice for Γ is (δ_{ij} is the Kronecker delta function)

$$\Gamma_{ij} = -\frac{\left(\frac{1}{2} + \frac{3}{4}C_0\right)\epsilon}{k_{sgs}} \delta_{ij} \qquad (6.11)$$

because this gives the correct rate of dissipation in decaying isotropic turbulence for inertia free particles. Fede et al. (2006) showed that this also holds in an LES context. In general, Γ must lead to relaxation of the modelled subgrid scales, i.e., must be negative definite.

Again, the model is closed in the sense that estimates for k_{sgs} and ϵ are needed.

6.3 Analytical assessment

Most authors of the models listed above present the performance of their model through results from Direct Numerical Simulation (DNS) and LES. Besides this data, there is little data available for model assessment.

An alternative method for model assessment is provided by an analytical approach, conducted in the present section. Here, no numerical simulation is involved. Instead, the stochastic moments A1 to A6 (cf. chapter 4) are computed analytically for each model. Comparison to the exact moments gives the model error.

Model assessment by analytical means in comparison to numerical simulation evidently bears the advantage that the results are not only valid for a specific configuration. Another advantage is that model errors in particle statistics can be related back to their sources, i.e., to specific terms in the model equation. This enables efficient model improvement. For example Reeks (2005) could clarify a discrepancy between two models for particles in RANS by a similar analysis.

On the other hand, an analytical approach is often based on several simplifications leading to systematic error sources. Therefore in section 6.4 additional results from numerical simulation are presented.

6.3.1 Framework for the analytical computations

For the analytical computations, two simplifications are made. One is that preferential concentration is neglected, i.e., it is assumed that the particles are distributed homogeneously in space. The second is that Stokes drag is assumed to be linear, i.e., (2.32) is replaced by

$$\frac{d\mathbf{u}_p}{dt} = \frac{1}{\tau_p}\left(\mathbf{u}_{f@p} - \mathbf{u}_p\right). \qquad (6.12)$$

Actually both simplifications are quite restrictive. Therefore the analysis is backed by numerical simulation where the original equation (2.32) was solved, cf. section 6.4.

However, the aim of this chapter is not to predict particle dynamics but to assess particle-

6 Presentation and assessment of existing particle-LES models

LES models. The results show that even for linear Stokes drag the models under consideration show significant defects. In configurations where Stokes drag is non-linear, these defects will show up as well. Thus, equation (6.12) is suited for the purpose of this work.

In the following, the models presented in section 6.2.2 are assessed by computing the statistical moments which result from the models and comparing these to their exact counterparts. The analysis focuses on the structure of the models and not the model parameters, i.e., for the Langevin-based models it is assumed that k_{sgs} and ϵ are correct and no restrictions for the model parameters C_0, Γ and T_L (such as equations (6.6) or (6.11)) are assumed. The question addressed is 'how good can a model be that is based on respective modelling strategy, assuming an optimal choice of model parameters'. Consequently, all model parameters are assumed to be independent of the modelled quantities.

6.3.2 First moments

In chapter 4, the first moments $\overline{\mathbf{x}_p}$, $\overline{\mathbf{u}_p}$ and $\overline{\mathbf{u}_f}$ were listed as assessment criteria. In the present section the first moments are computed for the solution one obtains when neglecting SGS effects and for the solutions which result from the three models under consideration. It is shown that all three models lead to error prone first moments.

First these errors are derived for each model respectively and then the errors are compared against each other in a spectral analysis further down in this section.

Section 4.2 showed that for the analysis of first moments it is sufficient to analyse the first moment of the fluid velocity seen by the particles $\overline{\mathbf{u}_{f@p}}$ because the other first moments ($\overline{\mathbf{u}_p}$ and $\overline{\mathbf{x}_p}$) are correct if and only if $\overline{\mathbf{u}_{f@p}}$ is correct.

First moments neglecting subgrid scale turbulence.

One might wonder whether SGS modelling for the particles is necessary at all. If no model is used, the transport equation for the first moment of the fluid velocity seen by the particle reads

$$\frac{\mathrm{d}\overline{u^N_{f@p,i}}}{\mathrm{d}t} = \frac{\mathrm{d}\overline{(\mathcal{G}u_{f,i})_{@p}}}{\mathrm{d}t} = \frac{\mathrm{d}\overline{u_{f@p,i}}}{\mathrm{d}t} - \frac{\mathrm{d}\overline{(\mathcal{H}u_{f,i})_{@p}}}{\mathrm{d}t}. \tag{6.13}$$

It should be noted that $\overline{(\mathcal{H}u_{f,i})_{@p}}$ is a Lagrangian average, i.e., it is not zero in an arbitrary flow field even if the filter is constructed such that the Eulerian average is zero.

Furthermore, for most LES models the implicitly defined filter is not a sharp spectral cutoff filter but affects all wavenumbers. Therefore the unresolved field $\mathcal{H}u_f$ can contain significant low-wavenumber contributions. Thus, the first moment of $\mathbf{u}_{f@p}$ is not predicted correctly if no model is used. Consequently also the first moments of $\mathbf{u}_p$ and $\mathbf{x}_p$ show deficiencies.

First moments using ADM.

With respect to the first moments, ADM performs better. The transport equation for $\overline{\mathbf{u}_{f@p}^{ADM}}$ reads

$$\frac{\mathrm{d}\overline{u_{f@p,i}^{ADM}}}{\mathrm{d}t} = \frac{\mathrm{d}\overline{u_{f@p,i}}}{\mathrm{d}t} - \frac{\mathrm{d}\overline{(\mathcal{H}^{N+1}u_{f,i})_{@p}}}{\mathrm{d}t}. \tag{6.14}$$

With respect to the infinity-norm, the error in the transport equation for the first moment is

$$e^{ADM} = \left\| \frac{\mathrm{d}\overline{(\mathcal{H}^{N+1}u_{f,i})_{@p}}}{\mathrm{d}t} \right\|_\infty. \tag{6.15}$$

This result was to be expected. It means that, if N is sufficiently large, then all scales which can be represented on the LES grid are resolved correctly whereas all scales smaller than the LES grid contribute to errors in the first moment.

For specific flows $\overline{\mathcal{H}^{N+1}\mathbf{u}_f}$ is negligible. Then the first moment is preserved in ADM. In general this cannot be assumed and the first moment is error prone.

First moments using the model of Shotorban & Mashayek.

Now the first moments of the model proposed by Shotorban & Mashayek are analysed. The first moment of $\mathbf{u}_{f@p}^{Sho}$ can be computed directly from equation (6.5),

$$\frac{\mathrm{d}\overline{u_{f@p,i}^{Sho}}}{\mathrm{d}t} = \frac{\mathrm{d}\overline{(\mathcal{G}u_{f,i})_{@p}}}{\mathrm{d}t} - \overline{u_{p,j}\left(\frac{\partial \mathcal{G}u_{f,i}}{\partial x_j}\right)_{@p}} + \overline{\left(\mathcal{G}\left(u_{f,j}\frac{\partial u_{f,i}}{\partial x_j}\right)\right)_{@p}} - \frac{\overline{u_{f@p,i}^{Sho}} - \overline{(\mathcal{G}u_{f,i})_{@p}}}{T_L}. \tag{6.16}$$

In the following it is shown that the first moment is not conserved with this model, i.e.,

$$\overline{\mathbf{u}_{f@p}^{Sho}} \neq \overline{\mathbf{u}_{f@p}}. \tag{6.17}$$

If the contrary would be true, i.e., $\overline{\mathbf{u}_{f@p}^{Sho}} = \overline{\mathbf{u}_{f@p}}$, then also $\mathrm{d}\overline{\mathbf{u}_{f@p}^{Sho}} / \mathrm{d}t = \mathrm{d}\overline{\mathbf{u}_{f@p}} / \mathrm{d}t$. This is equivalent to

$$\frac{\mathrm{d}\overline{u_{f@p,i}^{Sho}}}{\mathrm{d}t} = \frac{\mathrm{d}\overline{(\mathcal{G}u_{f,i})_{@p}}}{\mathrm{d}t} + \frac{\mathrm{d}\overline{(\mathcal{H}u_{f,i})_{@p}}}{\mathrm{d}t}, \tag{6.18}$$

thus

$$\overline{\left(\mathcal{H}\left(\frac{\partial u_{f,i}}{\partial t} + u_{p,j}\frac{\partial u_{f,i}}{\partial x_j}\right)\right)_{@p}} + \frac{\overline{u_{f@p,i}^{Sho}} - \overline{(\mathcal{G}u_{f,i})_{@p}}}{T_L} = \overline{\left(\mathcal{G}\left(u_{f,j}\frac{\partial u_{f,i}}{\partial x_j} - u_{p,j}\frac{\partial u_{f,i}}{\partial x_j}\right)\right)_{@p}}. \tag{6.19}$$

The left hand side of equation (6.19) is a high-frequency signal, the right hand side is at low frequency. Thus, equation (6.19) holds for an arbitrary flow field $\mathbf{u}_f$ only if both sides are zero.

6 Presentation and assessment of existing particle-LES models

For $\tau_p = 0$, particle velocity equals fluid velocity and the right hand side vanishes. This means that the low-frequency components of $\mathbf{u}_p$ are exact for $\tau_p = 0$. Gicquel et al. (2002) showed this in an FDF context. For $\tau_p > 0$, $\mathbf{u}_p \neq \mathbf{u}_{f@p}$ and the right hand side of equation (6.19) is not zero.

Also the left hand side cannot be zero for all values of τ_p because $\mathbf{u}_p$ depends on particle relaxation time but all other terms do not. Shotorban & Mashayek chose T_L such that in decaying isotropic turbulence the model predicts the correct rate of dissipation for $\tau_p = 0$. This means that in the best case T_L is set such that the left hand side is zero for $\tau_p = 0$,

$$\overline{\left(\frac{\partial \mathcal{H} u_{f,i}}{\partial t} + u_{f@p,j}\frac{\partial \mathcal{H} u_{f,i}}{\partial x_j}\right)_{@p}} + \frac{\overline{u_{f@p,i}^{Sho}} - \overline{(\mathcal{G} u_{f,i})_{@p}}}{T_L} = 0. \tag{6.20}$$

Then the overall model error reads

$$e^{Sho} = \left\|\overline{\left(u_{p,j}\frac{\partial u_{f,i}}{\partial x_j} - u_{f@p,j}\frac{\partial u_{f,i}}{\partial x_j}\right)_{@p}}\right\|_{\infty}. \tag{6.21}$$

Evidently, if $\tau_p > 0$ then the error e^{Sho} is not zero for an arbitrary flow field $\mathbf{u}_f$. Thus, the model proposed by Shotorban & Mashayek does not predict the first moments correctly. Therefore equation (6.17) is true.

First moments using the model of Simonin *et al.*

For the model proposed by Simonin et al. (1993), the transport equation for the first moment of $\mathbf{u}_f^{Sim}$ reads

$$\frac{d\overline{u_{f@p,i}^{Sim}}}{dt} = \frac{d\overline{(\mathcal{G} u_{f,i})_{@p}}}{dt} - \overline{u_{f@p,j}^{Sim}\left(\frac{\partial \mathcal{G} u_{f,i}}{\partial x_j}\right)_{@p}} + \overline{\left(\frac{\partial \mathcal{G} u_{f,i} u_{f,j}}{\partial x_j}\right)_{@p}} + \Gamma_{ij}\overline{u_{f@p,j}^{Sim'}}. \tag{6.22}$$

With the same argumentation as above, the first moment is correct if and only if

$$\overline{\left(\mathcal{H}\left(\frac{\partial u_{f,i}}{\partial t} + u_{p,j}\frac{\partial u_{f,i}}{\partial x_j}\right)\right)_{@p}} - \Gamma_{ij}\overline{u_{f@p,j}^{Sim'}} = \overline{\left(\mathcal{G}\left(-u_{f@p,j}^{Sim}\frac{\partial u_{f,i}}{\partial x_j} + u_{f,j}\frac{\partial u_{f,i}}{\partial x_j}\right)\right)_{@p}}. \tag{6.23}$$

Again, equation (6.23) is sorted with respect to high- and low-frequency components and thus again, $\overline{\mathbf{u}_{f@p}^{Sim}} = \overline{\mathbf{u}_{f@p}}$ if and only if the left hand side and the right hand side of equation (6.23) are zero. Now, it's possible to distinguish between

- a best case estimation: If the model works well then $\mathbf{u}_{@p}^{Sim} \approx \mathbf{u}_{f@p}$ and the model error in the low-frequency components is negligible in comparison to the error from the high-frequency components.

  With the same argumentation as above, the error from high-frequency components cannot be zero for all Stokes numbers and in the best case Γ is set such that the left hand side is zero for $\tau_p = 0$,

$$\overline{\left(\frac{\partial \mathcal{H} u_{f,i}}{\partial t} + u_{f@p,j}\frac{\partial \mathcal{H} u_{f,i}}{\partial x_j}\right)_{@p}} - \Gamma_{ij}\overline{u_{f@p,j}^{Sim}}' = 0 \qquad (6.24)$$

This means that the model error reads

$$e^{Sim} = \left\|\overline{\left(u_{p,j}\frac{\partial \mathcal{H} u_{f,i}}{\partial x_j} - u_{f@p,j}\frac{\partial \mathcal{H} u_{f,i}}{\partial x_j}\right)_{@p}}\right\|_\infty. \qquad (6.25)$$

- a worst case estimation: In the worst case, the deviations between $\mathbf{u}_{f@p}^{Sim}$ and $\mathbf{u}_{f@p}$ are significant and the modelling error reads

$$e^{Sim} = \left\|\overline{\left(\frac{\partial \mathcal{H} u_{f,i}}{\partial t} + u_{p,j}\frac{\partial \mathcal{H} u_{f,i}}{\partial x_j}\right)_{@p}} - \Gamma_{ij}\overline{u_{f@p,j}^{Sim}}'\right\|_\infty$$
$$+ \left\|\overline{\left(u_{f@p,j}\frac{\partial \mathcal{G} u_{f,i}}{\partial x_j} - u_{f@p,j}^{Sim}\frac{\partial \mathcal{G} u_{f,i}}{\partial x_j}\right)_{@p}}\right\|_\infty \qquad (6.26)$$

Concluding, the first moment is error prone for all models considered. However, for specific flows and/or specific particles, these errors can be small. This is discussed in the following section.

Spectral analysis of error terms.

In the following the error terms derived so far are discussed by spectral analysis. Two questions are addressed. The first is Stokes number dependence of the error. The discussion shows that in general for small Stokes number the Langevin-based models show less error in the first moment than ADM whereas for high Stokes number ADM performs better.

The second question concerns dependence between error in first moments and the flow structure. It is analysed to which extend eddies of a specific size affect the error in the first moment. For ADM it is clear that all eddies smaller than the LES grid are not reconstructed and are therefore error sources for $\mathbf{u}_{f@p}^{ADM}$. For the Langevin-based models it is shown that, given two eddies of different size but same kinetic energy, the smaller eddy has a greater effect on the first moment error.

Equations (6.21) and (6.25) show that for both Langevin-based models the first moments are exact for inertia free particles but not if $\mathbf{u}_p \neq \mathbf{u}_{f@p}$, assuming the presented best case estimation for the model by Simonin et al. The error terms e^{Sho} and e^{Sim} are convective terms stemming from the fact that in the models, the (SGS) fluid velocity seen by the particles is convected by fluid velocity instead of particle velocity. Especially for the model by Simonin et al. this defect cannot be circumvented easily because convection of SGS velocity involves the unclosed SGS stress tensor.

However, in the following it is shown that the error in the first moments can be small if either the particle velocity is not very different to fluid velocity or if the fluid velocity is very smooth such that $\mathbf{u}_f(\mathbf{x}_p(t_0)) + \mathbf{u}_{f@p}\Delta t) \approx \mathbf{u}_f(\mathbf{x}_p(t_0)) + \mathbf{u}_p\Delta t)$.

6 Presentation and assessment of existing particle-LES models

According to equations (6.21) and (6.25), the relative averaged error from the convective term reads in a best case estimation

$$e_i^{rel} = \frac{\overline{(u_{p,j} - u_{f@p,j})\left(\frac{\partial v_i}{\partial x_j}\right)_{@p}}}{\overline{u_{p,k}\left(\frac{\partial v_i}{\partial x_k}\right)_{@p}}}. \tag{6.27}$$

For the model by Shotorban & Mashayek, $\mathbf{v}$ equals $\mathbf{u}_f$ and for the model by Simonin *et al.*, $\mathbf{v}$ equals $\mathcal{H}\mathbf{u}_f$.

In order to analyse the effects of the carrier flow's structure, it is useful to conduct a spectral analysis. First, define the field

$$\hat{e}_i^{rel}(\mathbf{k}) = \frac{(u_{p,j} - u_{f@p,j})\, k_j \mathcal{FT}(v_i)(\mathbf{k})}{\overline{u_{p,k}\left(\frac{\partial v_i}{\partial x_k}\right)_{@p}}}. \tag{6.28}$$

It holds

$$e_i^{rel} = \overline{\left(\mathcal{FT}^{-1}\left(\hat{e}_i^{rel}(\mathbf{k})\right)\right)_{@p}}. \tag{6.29}$$

$\hat{e}_i^{rel}$ can be regarded as the Fourier transform of the (non-averaged) model error. Now, define a wavenumber dependent function $e_i(\|\mathbf{k}\|)$ which links the spectrum of $\mathbf{v}$ to the error $\hat{e}_i^{rel}$,

$$\hat{e}_i^{rel}(\mathbf{k}) \leq e_i(\|\mathbf{k}\|)\, \frac{|\mathcal{FT}(v_i)(\mathbf{k})|}{v_{rms}} \tag{6.30a}$$

$$e_i(\kappa) = \frac{\|\mathbf{u}_p - \mathbf{u}_{f@p}\|\, v_{rms}}{\overline{u_{p,k}\left(\frac{\partial v_i}{\partial x_k}\right)_{@p}}}\kappa \quad \text{with} \quad v_{rms} = \sqrt{\overline{v_i^2}}. \tag{6.30b}$$

The inequality is a result of the Cauchy-Schwarz theorem. The denominator is not Fourier transformed because the denominator only serves as a normalisation factor for the error analysis.

$\mathbf{e}$ expresses the sensitivity of the error to the scales present in the flow field $\mathbf{v}$ in dependence of $\mathbf{u}_p$. Now, this dependence is reformulated in terms of Stokes number.

Dependence of e on τ_p. The denominator of e_i can be transformed to

$$\overline{u_{p,j}\left(\frac{\partial v_i}{\partial x_j}\right)_{@p}} = u_{p,rms} v_{rms} \frac{\overline{u_{p,j}\left(\frac{\partial v_i}{\partial x_j}\right)_{@p}}}{u_{p,rms} v_{rms}} = u_{p,rms} v_{rms} c_i \tag{6.31}$$

where $u_{p,rms}$ denotes the particle's rms velocity

$$u_{p,rms}^2 = \overline{u_{p,i}^2}. \tag{6.32}$$

The term $c_i = \overline{\dfrac{u_{p,j}}{u_{p,rms}v_{rms}}\left(\dfrac{\partial v_i}{\partial x_j}\right)_{@p}}$ is the scaled covariance between particle velocity and gradient of $\mathbf{v}$. $\mathbf{v}$ is either the fluid velocity $\mathbf{u}_f$ (for the model proposed by Shotorban & Mashayek) or the SGS fluid velocity $\mathcal{H}\mathbf{u}_f$ (for the model proposed by Simonin et al.). Thus, for $\tau_p = 0$, $\mathbf{c}$ is the averaged convective term from the Navier–Stokes equation for the (SGS) fluid velocity. Therefore, $\mathbf{c}$ is in general not 0 for $\tau_p = 0$. For $\tau_p \to \infty$ the particles are not affected by the fluid. The direction of the particle velocity $u_{p,j}/u_{p,rms}$ is not correlated with the gradient of the fluid velocity and thus $\mathbf{c}(\tau_p = \infty) = 0$. Therefore, in a first estimate, $\mathbf{c}$ can be assumed to decrease with particle relaxation time.

In order to obtain a relation between $\mathbf{e}$ and τ_p, the following two results from Tchen's theory (see Hinze, 1975) are employed:

$$u_{p,rms}^2 = \overline{u_{p,i}^2} = \overline{u_{f@p,i}^2}\frac{t_{u@p}}{t_{u@p} + \tau_p} \tag{6.33}$$

and

$$\overline{(u_{p,i} - u_{f@p,i})^2} = \overline{u_{f@p,i}^2}\frac{\tau_p}{t_{u@p} + \tau_p}. \tag{6.34}$$

$t_{u@p}$ denotes the integral time scale of the fluid seen by the particle. Tchen originally developed his theory for isotropic turbulence but by appropriate substitution of $t_{u@p}$, formulas (6.33) and (6.34) can be extended to arbitrary flow configurations (cf. Hinze, 1975; Wang & Stock, 1993; Issa & Oliveira, 1997).

With equations (6.30b) to (6.34) a scaling law for the rms value of $\mathbf{e}$ can be deducted:

$$e_i^{rms}(\kappa) = \sqrt{\overline{e_i^2}(\kappa)} = \frac{\sqrt{\tau_p}}{\sqrt{t_{u@p}}c_i}\kappa. \tag{6.35}$$

If the flow field is known then equation (6.35) gives together with equation (6.30a) an estimate for the error in the first moment for the stochastic models in terms of $\mathbf{c}$.

The function $k_{i,e}(\tau_p) = \tau_p^{-1/2}c_i$ is an isoline of e_i^{rms}, i.e., $e_i^{rms}(k_{i,e}) = \text{const}$. This function is depicted in figure 6.1. Along this isoline, the contribution of an eddy to the error in the first moments is constant for the Langevin-based models. This means for the Langevin-based models that the error increases with particle relaxation time. The reader is reminded that this refers to the error in the first moment of the fluid velocity seen by the particles and not to the error in the particle velocity.

In figure 6.1, an arbitrary positioned highest resolved wavenumber (cutoff wavenumber) κ_c is denoted. The effect of ADM is to reduce the defect at wavenumbers smaller than κ_c, independent of the Stokes number. Therefore with ADM, $e_i^{rms}(\kappa)$ is quasi zero for $\kappa < \kappa_c$ and large for $\kappa > \kappa_c$.

Figure 6.1 shows that for each stochastic model, a critical particle relaxation time τ_c can be defined such that ADM is more accurate for $\tau_p > \tau_c$ and the stochastic model is more accurate for $\tau_p < \tau_c$. Of course, τ_c depends on the flow as well.

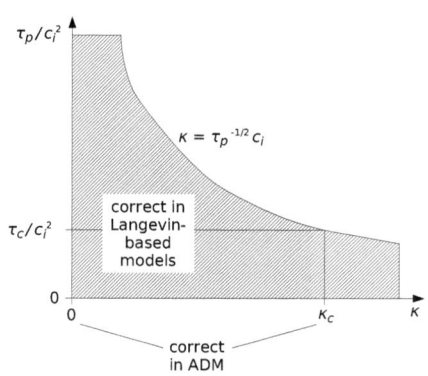
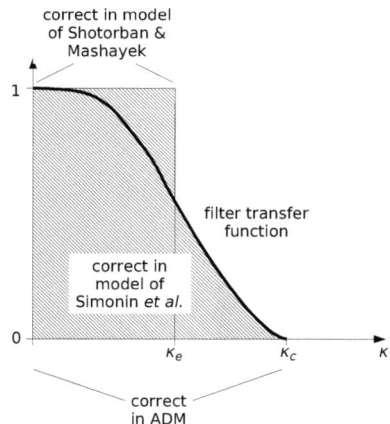

Figure 6.1: Sketch for modelling errors dependent on particle relaxation time and wavenumber. 'Correct' means that the corresponding scales are transported correctly, for all other scales the modelled convection is error prone.

Figure 6.2: Sketch for modelling errors for $\tau_p > \tau_c$. The filter transfer function is arbitrary. Again, 'correct' means that the corresponding scales are transported correctly, for all other scales the modelled convection is error prone.

Comparison of the two stochastic models against each other. Now the two stochastic models are compared against each other. In the previous analysis, $\mathbf{v}$ stands for $\mathbf{u}_f$ and $(\mathcal{I} - \mathcal{G})\mathbf{u}_f$ for the model of Shotorban & Mashayek and Simonin et al., respectively. The relation between the corresponding Fourier transformed velocities is expressed by the transfer function $\mathcal{FT}(G)$ of the LES filter,

$$\int_{\|\mathbf{k}'\|=\|\mathbf{k}\|} \|\mathcal{FT}(\mathcal{G}\mathbf{u}_f)(\mathbf{k}')\|^2 \, d\mathbf{k}' = |\mathcal{FT}(G)(\|\mathbf{k}\|)|^2 \int_{\|\mathbf{k}'\|=\|\mathbf{k}\|} \|\mathcal{FT}(\mathbf{u}_f)(\mathbf{k}')\|^2 \, d\mathbf{k}'. \qquad (6.36)$$

$\mathcal{FT}(G)$ is sketched in figure 6.2. The area below $\mathcal{FT}(G)$ corresponds to resolved eddies, the area above $\mathcal{FT}(G)$ corresponds to unresolved eddies.

Deficiencies of the model of Simonin et al. can be assigned to the unresolved velocity $\mathcal{H}\mathbf{u}_f$, i.e., the area above the transfer function, cf. figure 6.2. Furthermore, figure 6.1 allows to restrict this area to the region $\kappa > k_{i,e}$. For the model of Shotorban & Mashayek, all scales with wavenumbers greater than $k_{i,e}$ are not transported correctly. The filter transfer function does not come into play here.

Figure 6.2 reveals that, if the particle relaxation time is greater than τ_c, then the first moment error is smaller with the model proposed by Simonin et al. than for the model proposed by Shotorban & Mashayek. This gives a clue to the relation between the critical particle relaxation time of the two models. By definition, τ_c depends on the model error in the first moment. For the model proposed by Simonin et al., this error is smaller than for the model by Shotorban & Mashayek and thus the critical particle relaxation time is higher for Simonin et al.'s model,

$$\tau_c^{Sim} > \tau_c^{Sho}. \tag{6.37}$$

This means that, concerning first moments, at high particle relaxation time ($\tau_p > \tau_c^{Sim}$), ADM is the best choice considering accuracy of the first moment. For intermediate particle relaxation time $\tau_c^{Sim} > \tau_p > \tau_c^{Sho}$ the model proposed by Simonin et al. is advantageous towards ADM and for $\tau_p < \tau_c^{Sho}$, the model by Simonin et al. and the model proposed by Shotorban & Mashayek give acceptable results.

With these results, it is possible to select a problem-specific particle-LES model such that the errors in the first moments are minimised. If the first moments are all zero such as in isotropic turbulence, then of course all models analysed predict them exactly. In this case, the second moments come into play.

6.3.3 Second moments

In this section, it is shown that with ADM the second moment of particle velocity is only predicted correctly for high particle relaxation time and that the second moment in particle position is not correct, independent of particle relaxation time. For the stochastic models, it is shown that for low particle relaxation time (where first moments show little error), the error in the second moments is small.

In section 4.2 it was already stated that for the analysis of second moments, it suffices to analyse $u_{f@p,i}(\tau)u_{f@p,j}(t)$ for arbitrary t and $\tau \leq t$. Furthermore, for the analysis of second moments it can be assumed that errors in the first moments are negligible because only in this case errors in the second moments are relevant.

Thus, for the following analysis of the second moments, it is assumed that a small number $\delta > 0$ exists such that

- for the model of Shotorban & Mashayek

$$e^{Sho} < \delta \quad \Rightarrow \quad \left\| \overline{\left((u_{f,j} - u_{p,j}) \frac{\partial u_{f,k}}{\partial x_j} \right)_{@p}} \right\|_\infty < \delta, \tag{6.38}$$

- for the model of Simonin et al.

$$e^{Sim} < \delta \quad \Rightarrow \quad \left\| \overline{\left((u_{f,j} - u_{f@p,j}^{Sim}) \frac{\partial u_{f,i}}{\partial x_j} \right)_{@p}} \right\|_\infty \leq \delta, \tag{6.39}$$

- for ADM

$$N \to \infty. \tag{6.40}$$

6 Presentation and assessment of existing particle-LES models

This is a necessary condition for $\|\overline{\mathbf{u}_p^{ADM}} - \overline{\mathbf{u}_p}\| < \delta$ if no statistical properties of the flow are known.

In the following it is shown that for the stochastic models the error in the second moments is at the order of δ if the conditions (6.38) and (6.39) are extended as follows:

- for the model of Shotorban & Mashayek

$$\left\|\overline{\left(\mathcal{G}\left((u_{f,j} - u_{p,j})\frac{\partial u_{f,k}}{\partial x_j}\right)\right)^2}_{@p}\right\|_\infty = O(\delta^2), \tag{6.41}$$

- for the model of Simonin et al.

$$\left\|\overline{\left(\mathcal{G}\left((u_{f,j} - u_{f@p,j}^{Sim})\frac{\partial u_{f,i}}{\partial x_j}\right)\right)^2}_{@p}\right\|_\infty = O(\delta^2). \tag{6.42}$$

Equations (6.41) and (6.42) are not necessary conditions for small errors in the first moment. For specific flows it is possible that (6.38) (resp. (6.39)) holds but (6.41) (resp. (6.42)) holds not. On the other hand, it was stated previously that for the stochastic models the first moment error is small if the particle relaxation time is small, independent of the flow. In that case, equations (6.41) and (6.42) hold as well. Thus, it can be stated that if the particle relaxation time number is small enough such that the error in the first moment is negligible for an arbitrary flow field then equations (6.41) and (6.42) hold.

Second moments using ADM.

For ADM with $N \to \infty$ (assumption (6.40)), the resolved velocity $\mathbf{u}_{f@p}^{ADM}$ does not correlate with the unresolved velocity $\mathcal{H}^{N+1}\mathbf{u}_{f@p}$, cf. equation (5.15). Thus, with ADM the second moment of the fluid velocity seen by the particles reads

$$\overline{u_{f@p,i}^{ADM} u_{f@p,i}^{ADM}} = \overline{u_{f@p,i} u_{f@p,i}} - \overline{(\mathcal{H}^{N+1} u_{f,i})_{@p} (\mathcal{H}^{N+1} u_{f,i})_{@p}}. \tag{6.43}$$

This means that, as expected, the second moment of the fluid velocity seen by the particles is underestimated by ADM for all Stokes numbers.

Much more interesting for applications is the second moment of particle velocity and particle position. Equation (4.7b) allows to express the autocovariance of $\mathbf{u}_p^{ADM}$ by

$$\overline{u_{p,i}^{ADM}(\tau) u_{p,i}^{ADM}(t)} = \overline{u_{p,i}(\tau) u_{p,i}(t)}$$
$$- \frac{1}{\tau_p^2} \int_{-\infty}^{t} \int_{-\infty}^{\tau} \overline{(\mathcal{H}^{N+1} u_{f,i})_{@p}(t_1)(\mathcal{H}^{N+1} u_{f,i})_{@p}(t_2)} e^{\frac{t_1+t_2-t-\tau}{\tau_p}} dt_1\, dt_2. \tag{6.44}$$

The last term is the model error. For $\tau_p \to 0$, the integrand is a Dirac function and the error reads $\overline{(\mathcal{H}^{N+1} u_{f,i})_{@p}(\tau)(\mathcal{H}^{N+1} u_{f,i})_{@p}(t)}$, i.e., for $\tau_p = 0$ the error is identical to the autocovariance of the high-frequency fluctuations. For $\tau_p \to \infty$ one might get the impression that the error is negligible,

$$\lim_{\tau_p \to \infty} \overline{u_{p,i}^{ADM}(\tau) u_{p,i}^{ADM}(t)} = \lim_{\tau_p \to \infty} \overline{u_{p,i}(\tau) u_{p,i}(t)}$$
$$- \lim_{\tau_p \to \infty} \frac{1}{\tau_p^2} \int_{-\infty}^{t} \int_{-\infty}^{\tau} \overline{(\mathcal{H}^{N+1} u_{f,i})_{@p}(t_1) (\mathcal{H}^{N+1} u_{f,i})_{@p}(t_2)} \, dt_1 \, dt_2$$
$$= \lim_{\tau_p \to \infty} \overline{u_{p,i}(\tau) u_{p,i}(t)}. \tag{6.45}$$

However, in terms of relative error this is not true:

$$\lim_{\tau_p \to \infty} \frac{\left| \overline{u_{p,i}^{ADM}(\tau) u_{p,i}^{ADM}(t)} - \overline{u_{p,i}(\tau) u_{p,i}(t)} \right|}{\left| \overline{u_{p,i}(\tau) u_{p,i}(t)} \right|} =$$
$$= \lim_{\tau_p \to \infty} \frac{\int_{-\infty}^{t} \int_{-\infty}^{\tau} \overline{(\mathcal{H}^{N+1} u_{f,i})_{@p}(t_1) (\mathcal{H}^{N+1} u_{f,i})_{@p}(t_2)} \, dt_1 \, dt_2}{\int_{-\infty}^{t} \int_{-\infty}^{\tau} \overline{u_{f@p,i}(t_1) u_{f@p,i}(t_2)} \, dt_1 \, dt_2} \tag{6.46}$$

This term is the (positive) covariance of the high-frequency fluctuations divided by the autocovariance of the fluid velocity. Thus, this term is not zero.

The second moment error can be analysed further in homogeneous isotropic turbulence. In this case, the easiest (and most common) assumption for the shape of the autocovariance of the unresolved scales is an exponential function (see e.g. Hinze, 1975),

$$\overline{(\mathcal{H}^{N+1} u_{f,i})_{@p}(s) (\mathcal{H}^{N+1} u_{f,i})_{@p}(0)} = \overline{(\mathcal{H}^{N+1} u_{f,i})_{@p}^2} e^{-\frac{s}{t_{hi}}}. \tag{6.47}$$

It should be pointed out that t_{hi} depends on particle relaxation time because due to the interaction of centrifugal forces and Stokes drag, particles tend to cluster even if the flow is homogeneous.

With equation (6.47), the autocovariance of $\mathbf{u}_p^{ADM}$ can be written as (see Wang & Stock, 1993)

$$\overline{u_{p,i}^{ADM}(\tau) u_{p,i}^{ADM}(t)} = \overline{u_{p,i}(\tau) u_{p,i}(t)} - \frac{\overline{(\mathcal{H}^{N+1} u_{f,i})_{@p}^2}}{t_{hi}^2 - \tau_p^2} \left(t_{hi}^2 e^{\frac{\tau-t}{t_{hi}}} - t_{hi} \tau_p \, e^{\frac{\tau-t}{\tau_p}} \right). \tag{6.48}$$

With $\tau = t$, equation (6.48) gives the second moment of particle velocity $\overline{u_{p,i}^{ADM} u_{p,i}^{ADM}}$,

$$\overline{u_{p,i}^{ADM}(t) u_{p,i}^{ADM}(t)} = \overline{u_{p,i}(t) u_{p,i}(t)} - \frac{\overline{(\mathcal{H}^{N+1} u_{f,i})_{@p}^2}}{t_{hi} + \tau_p} t_{hi}. \tag{6.49}$$

Thus, kinetic energy of the particles is underestimated by ADM and the error vanishes for $\tau_p \to \infty$.

For the second moment in particle position, one obtains from equations (4.5b) and (6.48)

$$\frac{\mathrm{d}\overline{x_{p,i}^{ADM}x_{p,i}^{ADM}}}{\mathrm{d}t} = \frac{\mathrm{d}\overline{\overline{x}_{p,i}\overline{x}_{p,i}}}{\mathrm{d}t} - 2t_{hi}\overline{(\mathcal{H}^{N+1}u_{f,i})_{@p}^2}. \tag{6.50}$$

As mentioned above, t_{hi} depends on particle relaxation time due to clustering. The same holds for $\overline{(\mathcal{H}^{N+1}u_{f,i})_{@p}^2}$. However, for all values of τ_p these quantities are greater than zero. This means that for all values of τ_p the rate of dispersion is not predicted correctly.

This might seem surprising because for the limit $\tau_p \to \infty$ the particles don't move and thus rate of dispersion is zero. Of course, ADM gives the correct result here. Equation (6.50) seems to state otherwise.

Equation (6.50) was derived from equation (6.48) by integration. In the present case computing the limit $\tau_p \to \infty$ of equation (6.50) corresponds to computing the limit of an integral. On the other hand arguing that particles with $\tau_p \to \infty$ possess velocity 0 and thus do not disperse corresponds to computing the integral of the limit. In the present case, these two expressions are not equivalent:

$$\lim_{\tau_p \to \infty} \int_{\tau=-\infty}^{t} \text{eq. (6.48) } \mathrm{d}\tau: \quad \lim_{\tau_p \to \infty} \int_{\tau=-\infty}^{t} \frac{\tau_p \, e^{\frac{\tau-t}{\tau_p}}}{t_{hi}^2 - \tau_p^2} \mathrm{d}\tau = 1 \tag{6.51a}$$

$$\int_{\tau=-\infty}^{t} \lim_{\tau_p \to \infty} \text{eq. (6.48) } \mathrm{d}\tau: \quad \int_{\tau=-\infty}^{t} \lim_{\tau_p \to \infty} \frac{\tau_p \, e^{\frac{\tau-t}{\tau_p}}}{t_{hi}^2 - \tau_p^2} \mathrm{d}\tau = 0 \tag{6.51b}$$

The physical mechanism behind this is that particles with high but finite Stokes number move slowly but are not decelerated by the flow. This means that their kinetic energy is very small but their integral time scale is very large. The rate of dispersion, being the product of both quantities, is therefore not necessarily small.

Second moments for the model of Shotorban & Mashayek.

Now the second moments for the model proposed by Shotorban & Mashayek are analysed using identity (4.7b). It is shown that, if conditions (6.38) and (6.41) hold, then the second moments show little error if the model parameters are set optimal.

It is useful to split up $\mathbf{u}_{f@p}^{Sho}$ in filtered velocity seen by the particle $(\mathcal{G}\mathbf{u}_f)_{@p}$ and modeled fluctuations $\mathbf{u}_{f@p}^{Sho'} = \mathbf{u}_{f@p}^{Sho} - (\mathcal{G}\mathbf{u}_f)_{@p}$. Then it is possible to rewrite the autocovariance of the fluid velocity seen by the particles as

$$\overline{u_{f@p,i}^{Sho}(\tau)u_{f@p,i}^{Sho}(t)} = \overline{(\mathcal{G}u_{f,i})_{@p}(\tau)(\mathcal{G}u_{f,i})_{@p}(t)} + \overline{(\mathcal{G}u_{f,i})_{@p}(\tau)u_{f@p,i}^{Sho'}(t)}$$
$$+ \overline{u_{f@p,i}^{Sho'}(\tau)(\mathcal{G}u_{f,i})_{@p}(t)} + \overline{u_{f@p,i}^{Sho'}(\tau)u_{f@p,i}^{Sho'}(t)}. \tag{6.52}$$

In the following the three rightmost terms will be analysed under the assumptions listed above, equations (6.38) and (6.41).

The differential equation for $\mathbf{u}_{f@p}^{Sho'}$ along a particle path reads

$$d u_{f@p,i}^{Sho'} = \underbrace{\left(\mathcal{G} \left(u_{f,j} \frac{\partial u_{f,i}}{\partial x_j} \right) - u_{p,j} \frac{\partial \mathcal{G} u_{f,i}}{\partial x_j} \right)_{@p}}_{:= f_i(t)} dt + \sqrt{C_0 \epsilon} \, dW_i - \frac{u_{f@p,i}^{Sho'}}{T_L} dt \qquad (6.53)$$

The function $\mathbf{f}$ characterises the influence of the model errors from the first moment on $\mathbf{u}_{f@p}^{Sho'}$. Equation (6.38) states that $\|\mathbf{\bar{f}}\|_\infty$ is small but nevertheless $\mathbf{f}$ can not be set to zero at this point because for the computation of the second moment $\overline{u_{p,i}^{Sho}(\tau) u_{p,i}^{Sho}(t)}$ the function $\mathbf{f}$ will be integrated and averaged and it must be shown that this term is small.

In order to separate the undesired effect of $\mathbf{f}$ from the desired effects of the model, define a stochastic process $\tilde{\mathbf{u}}$ by

$$\tilde{u}_i := \frac{1}{\sqrt{C_0 \epsilon}} \left(u_{f@p,i}^{Sho'} - g_i \right) \quad \text{with } g_i(t) = \int_{-\infty}^{t} f_i(\tau) e^{\frac{\tau - t}{T_L}} d\tau. \qquad (6.54)$$

One can verify by substitution that $\tilde{\mathbf{u}}$ is solution of the SDE

$$d\tilde{\mathbf{u}} = -\frac{\tilde{\mathbf{u}}}{T_L} dt + d\mathbf{W}. \qquad (6.55)$$

Thus, $\tilde{\mathbf{u}}$ is independent of $\mathcal{G}\mathbf{u}_f$, the first moment error $\mathbf{f}$ and its history $\mathbf{g}$.

Rewriting $\mathbf{u}_{f@p}^{Sho'}$ as $\mathbf{u}_{f@p}^{Sho'} = \sqrt{C_0 \epsilon} \tilde{\mathbf{u}} + \mathbf{g}$, one obtains

$$\overline{(\mathcal{G} u_{f,i})_{@p}(\tau) u_{f@p,i}^{Sho'}(t)} = \overline{(\mathcal{G} u_{f,i})_{@p}(\tau) g_i(t)} \qquad (6.56a)$$

$$\text{and} \qquad \overline{u_{f@p,i}^{Sho'}(\tau) u_{f@p,i}^{Sho'}(t)} = C_0 \epsilon \overline{\tilde{u}_i(\tau) \tilde{u}_i(t)} + \overline{g_i(\tau) g_i(t)}. \qquad (6.56b)$$

Equation (6.56a) can be reduced to $\overline{(\mathcal{G} u_{f,i})_{@p}(\tau) g_i(t)} = O(\delta)$ due to equations (6.54) and (6.41) and the Cauchy-Schwarz inequality. It should be noted that for $\tau_p = 0$, $\mathbf{u}_p = \mathbf{u}_{f@p}$ and $\delta = 0$. Finally, equation (6.55) gives $\overline{\tilde{u}_i(\tau) \tilde{u}_i(t)} = \frac{T_L}{2} e^{-\frac{|\tau - t|}{T_L}}$. Thus, equation (6.52) reads

$$\overline{u_{f@p,i}^{Sho}(\tau) u_{f@p,i}^{Sho}(t)} = \overline{(\mathcal{G} u_{f,i})_{@p}(\tau) (\mathcal{G} u_{f,i})_{@p}(t)} + \frac{C_0 \epsilon T_L}{2} e^{-\frac{|\tau - t|}{T_L}} + O(\delta) \qquad (6.57)$$

and equation (4.7b) gives

$$\overline{u_{p,i}^{Sho}(\tau) u_{p,i}(t)^{Sho}} = \frac{1}{\tau_p^2} \int_{-\infty}^{t} \int_{-\infty}^{\tau} \overline{(\mathcal{G} u_{f,i})_{@p}(t_1) (\mathcal{G} u_{f,i})_{@p}(t_2)} e^{\frac{t_1 + t_2 - t - \tau}{\tau_p}} dt_1 \, dt_2$$

$$+ \frac{C_0 \epsilon T_L^2}{2(\tau_p^2 - T_L^2)} \left(\tau_p \, e^{\frac{\tau - t}{\tau_p}} - T_L e^{\frac{\tau - t}{T_L}} \right) + O(\delta). \qquad (6.58)$$

The first term on the right hand side corresponds to $\mathbf{u}_p^{ADM}$ with $N = 0$. Equation (6.48) can

6 Presentation and assessment of existing particle-LES models

be derived for all values of N (of course, t_{hi} will depend on N) and gives

$$\overline{u_{p,i}^{Sho}(\tau)u_{p,i}(t)^{Sho}} = \overline{u_{p,i}(\tau)u_{p,i}(t)} - \frac{\overline{(\mathcal{H}u_{f,i})^2_{@p}}}{\tau_p^2 - t_{hi}^2}\left(t_{hi}\tau_p\, e^{\frac{\tau-t}{\tau_p}} - t_{hi}^2 e^{\frac{\tau-t}{t_{hi}}}\right)$$
$$+ \frac{C_0\epsilon T_L^2}{2(\tau_p^2 - T_L^2)}\left(\tau_p\, e^{\frac{\tau-t}{\tau_p}} - T_L e^{\frac{\tau-t}{T_L}}\right) + O(\delta). \tag{6.59}$$

For $\delta \to 0$ evidently the second moment is exact if

$$T_L = t_{hi} \quad \text{and} \quad C_0 = \frac{2\overline{(\mathcal{H}u_{f,i})^2_{@p}}}{\epsilon T_L} \quad \text{for all } i. \tag{6.60}$$

This is only possible if the SGS velocity is isotropic. In the context of PDF methods for forced isotropic turbulence a common choice is (see Pope, 2000)

$$T_L = \frac{\overline{\|\mathcal{H}u_f\|_2^2}}{\frac{3}{2}C_0\epsilon} = \frac{4k_{sgs}}{3C_0\epsilon}, \tag{6.61}$$

in accordance with equation (6.60).

For anisotropic flows appropriate extensions are possible (see Haworth & Pope, 1986). Thus, if the model parameters are set correctly, then the errors in the second moments are insignificant for the model of Shotorban & Mashayek if particle relaxation time is small enough such that δ from (6.38) is negligible.

Second moments using the model of Simonin *et al.*

For the model proposed by Simonin *et al.* the second moments can be computed analogous to the second moments for the model of Shotorban & Mashayek with

$$f_i(t) = -u_{f@p,j}^{Sim}\left(\frac{\partial \mathcal{G}u_{f,i}}{\partial x_j}\right)_{@p} + \left(\frac{\partial \mathcal{G}(u_{f,i}u_{f,j})}{\partial x_j}\right)_{@p}. \tag{6.62}$$

In the model of Simonin *et al.* anisotropy is directly included via the matrix Γ. This leads to more complicated terms for the second moments than for the model of Shotorban & Mashayek but the computational steps are similar. Again, due to equation (6.42) and $\overline{\tilde{u}_i(\tau)\tilde{u}_i(t)} = -\frac{1}{2}\left(e^{\Gamma|\tau-t|}\Gamma^{-1}\right)_{ii}$ the following equation can be derived:

$$\overline{u_{p,i}^{Sim}(\tau)u_{p,i}(t)^{Sim}} = \overline{u_{p,i}(\tau)u_{p,i}(t)} - \frac{\overline{(\mathcal{H}u_{f,i})^2_{@p}}}{\tau_p^2 - t_{hi}^2}\left(t_{hi}\tau_p\, e^{\frac{\tau-t}{\tau_p}} - t_{hi}^2 e^{\frac{\tau-t}{t_{hi}}}\right)$$
$$+ \frac{C_0\epsilon\left(\Gamma - \tau_p I\right)^{-1}}{4}\left(e^{\Gamma(t-\tau)} + e^{\frac{\tau-t}{\tau_p}}\right)$$
$$+ \frac{C_0\epsilon\left(\Gamma + \tau_p I\right)^{-1}}{4}\left(e^{\Gamma(t-\tau)} - e^{\frac{\tau-t}{\tau_p}}\right) + O(\delta). \tag{6.63}$$

If one substitutes $\Gamma = -1/T_L I$ in this equation, then one obtains equation (6.59). As mentioned previously, the model of Shotorban & Mashayek can be extended for anisotropic flows with the method proposed by Haworth & Pope (1986). Then, for the model of Shotorban & Mashayek one obtains equation (6.63) instead of equation (6.59). Therefore with the same argumentation as above, also for Simonin *et al.*'s model it can be stated that with the correct choice of model parameters the errors in the second moments are insignificant if particle relaxation time is small enough such that δ from (6.42) is negligible.

6.4 Numerical assessment

The present section contains results from a priori and a posteriori analysis for assessment of the three models presented in section 6.2.2. The section reviews published numerical results (section 6.4.1) and results from own simulations (sections 6.4.2 and 6.4.3). In all numerical simulations the particle transport equation (2.32) was solved.

For the present simulations, the testcase is isotropic turbulence at $Re_\lambda = 52$. Details on the simulation parameters can be found in chapter 5. The models were assessed on this testcase by a priori and posteriori analysis. For the a priori analysis, the DNS field was filtered in each time step by a top hat filter as described in chapter 5. In contrast to the method described in that chapter, in the present analysis the filtered field was sampled on a mesh which corresponds to the LES grid. This field was taken as a field comparable to a LES field.

The a priori analysis was conducted differently for ADM than for the stochastic models in order to account for the respective model assumptions. More details on this issue are given in the respective sections.

In isotropic turbulence, all first moments are zero. Of course all models recover this. Therefore only second moments are analysed numerically.

For each model, some parameters must be set and additional equations must be solved. This is explained in detail for each model before the respective results are presented.

6.4.1 Numerical results from literature

Published numerical results with ADM are in accordance with the results of the analytical assessment, section 6.3. Kuerten (2006b) analysed ADM in particle-laden turbulent channel flow. The Reynolds number based on friction velocity was $Re_\tau = 150$. He analysed ADM for particles with Stokes numbers of $St = 1$, 5 and 25 by a posteriori analysis. He found that ADM significantly improves rms values of the wall normal component of the particle velocity, indicating that the second moments are improved by ADM, in accordance with equation (6.49). Also in accordance with equation (6.49), Kuerten found greater improvement for high Stokes number than for low Stokes number. However, his analysis was restricted to low Reynolds number and consequently small filter widths because at that time no reference data for higher Reynolds number was available.

Shotorban & Mashayek (2005) conducted simulations of particles in a turbulent shear layer. Also in their simulations the Reynolds number is low ($Re_\lambda = 24$) and the filter width is small ($\kappa_{c,DNS} = 2\kappa_{c,LES}$). The Stokes numbers read $St = 0.35$, 1.2 and 4.6. The authors found that ADM improves particle dispersion and that the improvement is independent of Stokes number, in accordance with equation (6.50).

Also for the Langevin-based models, numerical results are available in literature. Shotorban & Mashayek (2005) analysed their model in decaying isotropic turbulence and found that for small Stokes numbers ($St \leq 2.5$) the model leads to correct second moments whereas at higher Stokes number significant deviations can be observed. In their paper they suspect that this might be due to the 'assumption that the velocity of the seen fluid particles evolve similar to that of the fluid particles'. This is in accordance with the present findings from the analytical computations. These show that, if $\|\overline{(\mathcal{G}((u_{f,j} - u_{p,j})\frac{\partial u_{f,k}}{\partial x_j}))_{@p}^2}\|_\infty$ is small (equation (6.41)), then the model predicts second moments correctly, equations (6.57) and (6.59).

Fede et al. (2006) analysed the model of Simonin et al. (1993) in forced isotropic turbulence and found that the model leads to correct kinetic energy for the particles. Also in their simulations, Stokes numbers only ranged up to $St = 5$. Therefore, Fede et al. (2006) could not observe the errors predicted in the present study for the second moment at high Stokes number.

Concluding, all published results from numerical simulation are in accordance with the present analytical estimates but this data is not satisfactory. For ADM, only data at small Reynolds numbers and small filter widths is available and for the Langevin-based models to date only data at small Stokes numbers is published. Furthermore, the data density over the Stokes number range is not satisfactory. Data rather correspond to probes at specific Stokes numbers but from this data no Stokes number dependent behaviour of the models can be deduced. Therefore in the present work additional simulations for analysis of these models were conducted.

6.4.2 Numerical assessment of ADM

Section 4.3 discussed already the various possibilities for an a priori analysis. In the present work, in the a priori analysis for ADM for each particle two different values for the particle velocity were computed simultaneously, cf. figure 4.1. One value, referred to as DNS particle velocity, is the velocity obtained from the DNS flow field and the particle transport equation (2.32). The second value, referred to as modelled particle velocity, is the velocity obtained from a filtered DNS field together with the respective model. The particles were tracked with the DNS particle velocity as in chapter 5 and statistical samples were taken from the modelled velocity. With this method the numerical results should correspond to the analytical results except for the non-linearity of the Stokes drag. For ADM, this approach is not contradictory with the model assumptions because the model does not explicitly include the small scale effects on the particle path.

There is no explicit parameter in the model but the filter needs to be presumed. This actually allows for a huge number of parameters.

In the present work, the ADM defiltering operator was computed in three different ways. First, it was computed as proposed by Kuerten (2006b). Second, it was computed making use of the DNS spectrum and third, a model spectrum was used.

The first approach was conducted in order to test Kuerten's approach directly. In this model, a box filter is presumed. As mentioned above, the present study focuses on the structure of the models. Therefore ADM was generalised by allowing more general presumed filters. This leads to the second and third approach.

In the following, all results from the first approach are labeled ADM[Kuerten], results from the second approach ADM^{DNS} and results from the third approach ADM^{mod}.

Kuerten's ADM approach

Kuerten (2006b) states that it is important to choose the defiltering operator such that it matches the fluid-LES model. He tested two fluid-LES models, namely ADM and the dynamic Smagorinsky model proposed by Germano et al. (1991). The latter model is very similar to the fluid-LES model implemented in this work, presented in section 3.1.2. As test filter, Kuerten implemented a box filter at twice the grid scale, like in the present work. Kuerten stresses that for deconvolution of the fluid velocity for the particles, the defiltering operator should be constructed from the same filter. He proposes to approximate the defiltering operator with a Taylor expansion up to second-order in the filter width. More precisely, this means that the filter is approximated in 1D by

$$\mathcal{G}u_f(x) = \frac{1}{\Delta} \int_{x-\frac{\Delta}{2}}^{x+\frac{\Delta}{2}} u_f(\xi) \, \mathrm{d}\xi$$

$$= \frac{1}{\Delta} \int_{x-\frac{\Delta}{2}}^{x+\frac{\Delta}{2}} \left(u_f(x) + \frac{\partial u_f}{\partial x}(\xi - x) + \frac{1}{2}\frac{\partial^2 u_f}{\partial x^2}(\xi - x)^2 + O(|\xi - x|^3) \right) \mathrm{d}\xi$$

$$= u_f(x) + \frac{\partial^2 u_f}{\partial x^2}\frac{\Delta^2}{24} + O(\Delta^3) \qquad (6.64)$$

where Δ denotes the cell width of the LES. Then, one truncates the $O(\Delta^3)$-term and the operator $\mathcal{H} = \mathcal{I} - \mathcal{G}$ simplifies to

$$\mathcal{H}u_f(x) = -\frac{\partial^2 u_f}{\partial x^2}\frac{\Delta^2}{24}. \qquad (6.65)$$

For $n > 1$, $\mathcal{H}^n u_f$ consists only of terms of order $O(\Delta^3)$, for example

$$\mathcal{H}^2 u_f(x) = -\frac{\partial^2 \mathcal{H}u_f}{\partial x^2}\frac{\Delta^2}{24} = \frac{\partial^4 u_f}{\partial x^4}\frac{\Delta^4}{24^2}. \qquad (6.66)$$

Therefore, truncation of all $O(\Delta^3)$-terms yields for the defiltered velocity (cf. equation (6.4))

$$u_f^{ADM} = \sum_{n=0}^{N} \mathcal{H}^n \mathcal{G} u_f = \mathcal{G} u_f - \frac{\partial^2 \mathcal{G} u_f}{\partial x^2}\frac{\Delta^2}{24}. \qquad (6.67)$$

It should be noted that this effectively means $N = 1$, as remarked by Kuerten (2006b) himself. With a second-order disretisation of the second derivative on an equidistant grid

6 Presentation and assessment of existing particle-LES models

$x_i = i$ one obtains

$$u_f^{ADM}(x_i) = \mathcal{G}u_f(x_i) - \frac{\mathcal{G}u_f(x_{i+1}) + \mathcal{G}u_f(x_{i-1}) - 2\mathcal{G}u_f(x_i)}{\Delta^2}\frac{\Delta^2}{24}$$
$$= \frac{13}{12}\mathcal{G}u_f(x_{i-1}) - \frac{1}{24}\mathcal{G}u_f(x_i) - \frac{1}{24}\mathcal{G}u_f(x_{i+1}). \tag{6.68}$$

For extension to three dimensions, the dimensional splitting technique is implemented as proposed by Stolz et al. (2001a). For an equidistant Cartesian grid $x_i = i, y_j = j, z_k = k$ this technique reads

$$\mathbf{u}_f^{ADM}(i,j,k) = \sum_{i_0=-\infty}^{\infty}\sum_{j_0=-\infty}^{\infty}\sum_{k_0=-\infty}^{\infty} w_{i_0,j_0,k_0}^{ADM,3D}\mathcal{G}u_f(i+i_0, j+j_0, k+k_0) \tag{6.69a}$$

$$w_{i_0,j_0,k_0}^{ADM,3D} = w_{i_0}^{ADM}w_{j_0}^{ADM}w_{k_0}^{ADM}. \tag{6.69b}$$

Thus, the three-dimensional stencil $w^{ADM,3D}$ is the product of three one-dimensional stencils w^{ADM}. For Kuerten's stencil, equation (6.68) defines w^{ADM} to

$$w_{i_0}^{ADM} = \begin{cases} \frac{13}{12} & \text{for } i_0 = 0 \\ -\frac{1}{24} & \text{for } i_0 = \pm 1 \\ 0 & \text{else.} \end{cases} \tag{6.70}$$

This stencil is plotted in figure 6.6. The corresponding transfer function $\mathcal{FT}(w^{ADM})$ is shown in figure 6.7. In the following this approach is referred to as ADM^{Kuerten}.

ADM approach via spectra

In addition to Kuerten's approach, two different ADM stencils were implemented. For both stencils, the presumed filter is computed by comparison of the LES spectrum against target spectra.

The overall approach is somewhat similar to the approach of Stolz et al. (2001a). In order to reduce numerical complexity, they restricted the ADM stencil to a maximum of 5 LES cells in each direction, i.e., the stencil covers $5^3 = 125$ cells. It should be mentioned that this means higher computational costs than the ADM^{Kuerten} approach which covers only $3^3 = 27$ cells.

Furthermore, Stolz et al. (2001a) implemented the dimensional splitting technique mentioned above. The ADM stencil is then defined by the five-dimensional vector $\mathbf{w}^{ADM}$ (subscripts running from -2 to 2).

Stolz et al. (2001a) considered that for inhomogeneous flow it is desirable to use a defiltering operator with non-uniform filter width. For such filters, filtering and differentiation does not commute, a so called commutation error arises. This is not desirable because the LES equations were derived under the condition that filtering and differentiation commutes. Therefore Stolz et al. (2001a) required $\mathbf{w}^{ADM}$ to be such that the commutation error is of third-order in terms of cell width. Furthermore they required constants to remain constant after defiltering. Among the possible defilter operators they chose the one which is as symmetric as possible.

In the present work the commutation error is not an issue because all numerical experiments were conducted in isotropic turbulence. Therefore the restrictions imposed on the defiltering operator are slightly different to those proposed by Stolz et al. (2001a). In accordance to Stolz et al. (2001a), the stencil spans also 125 cells using the dimensional splitting technique and constants are required to remain constant after defiltering. In contrast to Stolz et al. (2001a), the stencil is completely symmetric. Additionally, a target spectrum E^{target} is imposed. This spectrum predicts a specific kinetic energy below the cutoff wavenumber $\int_0^{\kappa_c} E^{target}(\kappa) \, d\kappa$. The defiltering operator is required to produce this kinetic energy. Among these operators one is selected by optimisation such that the defiltered spectrum below the cutoff wavenumber is as close as possible to the target spectrum, i.e., $\mathbf{w}^{ADM}$ is solution to

$$\int_0^{\kappa_c} \left(|\mathcal{FT}(\mathbf{w})(\kappa)|^2 \, E^{LES}(\kappa) - E^{target}(\kappa) \right)^2 \, d\kappa \to \min \tag{6.71a}$$

$$w_{-1} = w_1 \tag{6.71b}$$

$$w_{-2} = w_2 \tag{6.71c}$$

$$\sum_{j=-2}^{2} w_j = 1 \tag{6.71d}$$

$$\int_0^{\kappa_c} |\mathcal{FT}(\mathbf{w})(\kappa)|^2 \, E^{LES}(\kappa) \, d\kappa = \int_0^{\kappa_c} E^{target}(\kappa) \, d\kappa. \tag{6.71e}$$

$\mathcal{FT}(\mathbf{w})(\kappa) = \sum_{j=-2}^{2} w_j e^{-i\kappa j}$ denotes the continuous Fourier transform of the discrete vector $\mathbf{w} \in \mathbf{R}^5$. Condition (6.71d) guarantees that constants remain constant after deconvolution and condition (6.71e) provides for correct large scale kinetic energy.

E^{LES} is the LES spectrum. The idea is to enhance the spectrum below the LES cutoff wavenumber κ_c such that $|\mathcal{FT}(\mathbf{w})(\kappa)|^2 E^{LES}(\kappa)$ approaches the target spectrum. Following this logic, for the a priori analysis E^{LES} should be replaced by the spectrum of the filtered DNS velocity. On the other hand, figure 5.4 shows that for the present testcase (isotropic turbulence at $Re_\lambda = 52$), the spectrum of the filtered field and the LES spectrum do not differ significantly. Therefore also for the a priori analysis the LES spectrum was used for construction of the ADM stencil $\mathbf{w}$, facilitating comparison of results from a priori and a posteriori analysis.

In the following, two choices for E^{target} are taken, namely first, E^{target} is replaced by the spectrum computed by DNS and second, E^{target} is replaced by Pope's model spectrum from section 2.1.3. The first approach is referred to as ADM^{DNS}, the second as ADM^{mod}. In both cases equations (6.71a) to (6.71e) were solved numerically.

The second approach, i.e., using a model spectrum as target spectrum, is important because for a realistic LES, there is usually no DNS spectrum available. On the other hand, this approach includes errors from the model spectrum. This spectrum is shown in figure 6.3 together with the DNS spectrum. Evidently the model spectrum shows a larger inertial subrange than the DNS spectrum, the $\kappa^{-5/3}$-regime extends towards higher wavenumbers. Although the differences between the two spectra are not negligible, this

6 Presentation and assessment of existing particle-LES models 101

model spectrum was used because it is known that for high Reynolds numbers, the model spectrum fits very well with experimentally and numerically computed spectra (cf. Pope, 2000).

Taking the DNS spectrum as target spectrum (ADM^{DNS}), one obtains for the isotropic turbulence test case at $Re_\lambda = 52$ (for details on this test case cf. chapter 5), the stencil $\mathbf{w}^{ADM}$ depicted in figure 6.4 and the corresponding transfer function $\mathcal{FT}(\mathbf{w}^{ADM})$, shown in figure 6.5 (continuous line). The dashed line of figure 6.5 shows $\sqrt{E^{LES}/E^{DNS}}$. Its inverse, $\sqrt{E^{DNS}/E^{LES}}$ is the target function for $\mathcal{FT}(\mathbf{w}^{ADM})$. $\mathcal{FT}(\mathbf{w}^{ADM})$ cannot be identical to

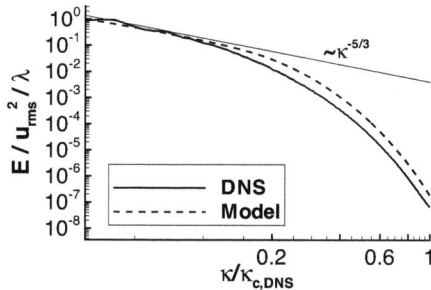

Figure 6.3: Spectrum from DNS and model spectrum propsed by Pope (2000) for $Re_\lambda = 52$.

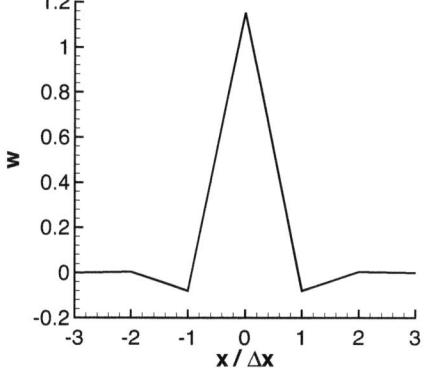

Figure 6.4: ADM stencil for $Re_\lambda = 52$ obtained by optimisation against DNS spectrum (ADM^{DNS}).

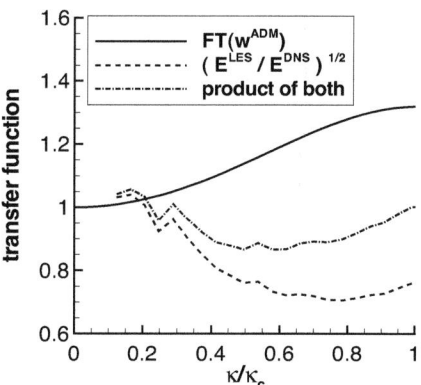

Figure 6.5: Continuous line: Transfer function of ADM stencil shown in figure 6.4, dashed line: LES transfer function $\sqrt{E^{LES}/E^{DNS}}$ obtained a posteriori from DNS and LES of isotropic turbulence at $Re_\lambda = 52$, dash-dotted line: product of both transfer functions.

$\sqrt{E^{DNS}/E^{LES}}$ because the stencil's support is restricted.

The dash-dotted line represents $|\mathcal{FT}(\mathbf{w}^{ADM})|\sqrt{E^{LES}/E^{DNS}}$. The target function (6.71a) of the optimisation aims at $|\mathcal{FT}(\mathbf{w}^{ADM})|\sqrt{E^{LES}/E^{DNS}} \equiv 1$.

With ADM^{mod}, one obtains the stencil shown in figure 6.6 and the transfer function shown in figure 6.7. Evidently this approach leads to a very much stronger amplification around κ_c than if one takes the DNS spectrum as target spectrum. This was to be expected because around κ_c the model spectrum is higher than the DNS spectrum, cf. figure 3.2.

Some more results on ADM^{DNS} and ADM^{mod} can be found in section 7.1.2.

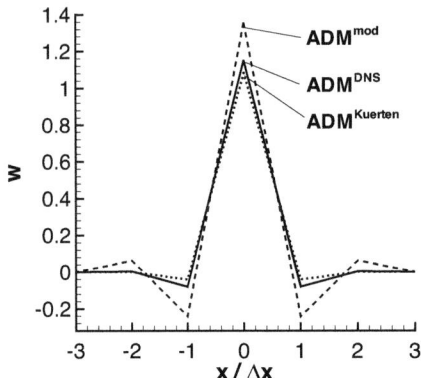
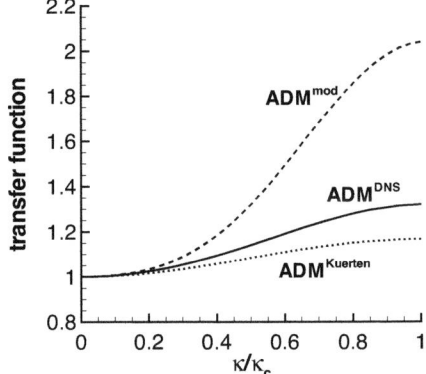

Figure 6.6: ADM stencil for $Re_\lambda = 52$ obtained by Kuerten's approach (dotted line), by optimisation against the DNS spectrum (continuous line) and by optimisation against Pope's model spectrum, cf. section 2.1.3 (dashed line).

Figure 6.7: Transfer function of ADM stencils shown in figure 6.6.

Second moments computed with ADM

As mentioned above, in isotropic turbulence the first moments are predicted exactly for all models. Therefore numerical results concerning second moments follow.

Figures 6.8 to 6.10 show second moments of the velocity seen by the particles, particle velocity and particle position for all three ADM stencils. In addition, results from filtered DNS and LES without particle-LES model are shown. In order to obtain comparable results, the presented data from filtered DNS without model corresponds to the data which ADM receives, i.e., in particular in the filtered DNS the particles were traced along unfiltered paths and the filtered data was sampled on the LES grid, cf. section 4.3. In contrast, the filtered DNS results from chapter 5 were obtained by filtering on the DNS grid. This means that the filtered DNS results from this section contain a higher interpolation error than the filtered

DNS results from chapter 5.

The kinetic energy seen by the particles (figure 6.8) shows qualitatively the same effects which were already observed when neglecting subgrid fluctuations, cf. section 5.5.1. Also results from a priori and a posteriori analysis are qualitatively equal. All ADM solutions show a shift along the Stokes axis in comparison to DNS. The reason for that was already explained in section 5.5.1. Furthermore, ADM results show smaller kinetic energy seen by the particles than DNS, in accordance with equation (6.43). Kuerten's model shows strongest deviations from DNS, the other two ADM stencils recover $k_{u@p}$ significantly better. This is not surprising because Kuerten's approach corresponds to a single defiltering step, $N = 1$, cf. equations (6.65) to (6.67). The other two approaches must perform better here because the wider stencils allow for larger values of N, leading to higher kinetic energy.

At first sight, the comparison of ADM^{DNS} against ADM^{mod} is surprising. ADM^{mod} shows closer resemblance to the unfiltered result than ADM^{DNS} although ADM^{DNS} is based on the unfiltered field. Actually this is an effect of two errors cancelling out each other. Around the cutoff wavenumber the model spectrum is higher than the DNS spectrum but interpolation leads to strong damping around the cutoff wavenumber. In other words, for ADM^{mod} the damping properties of the interpolation scheme brings the spectrum seen by the particles closer to the spectrum of the DNS flow field. More details on that issue can be found in section 7.1.2.

Figure 6.9 shows k_u, the kinetic energy of the particles themselves. In accordance with equation (6.49), k_u is underestimated by ADM and with $St \to \infty$ the error vanishes.

Most interesting is the rate of dispersion, shown in figure 6.10. Here, results from a priori and a posteriori analysis differ *qualitatively*. In accordance with the analytical predictions, equation (6.50), the a priori analysis shows that ADM leads to an underprediction of the rate of dispersion for all Stokes numbers.

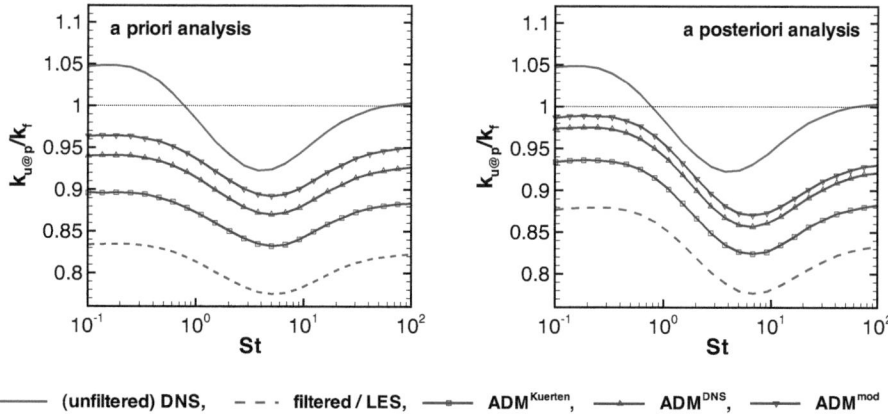

Figure 6.8: A priori (left) and a posteriori (right) analysis of ADM, kinetic energy seen by particles (second moment in fluid velocity seen by particles).

The a posteriori analysis shows that actually in LES the rate of dispersion is overestimated by ADM. The qualitative difference between a priori analysis and a posteriori analysis is due to a defect in the fluid-LES model, cf. section 5.5.2. The fluid-LES model leads to too high life times for the large eddies and thus to an overprediction of particle dispersion.

Concluding, all results are in accordance with the previously presented analytical computations. The kinetic energy of the particles is, especially for small Stokes number, under-

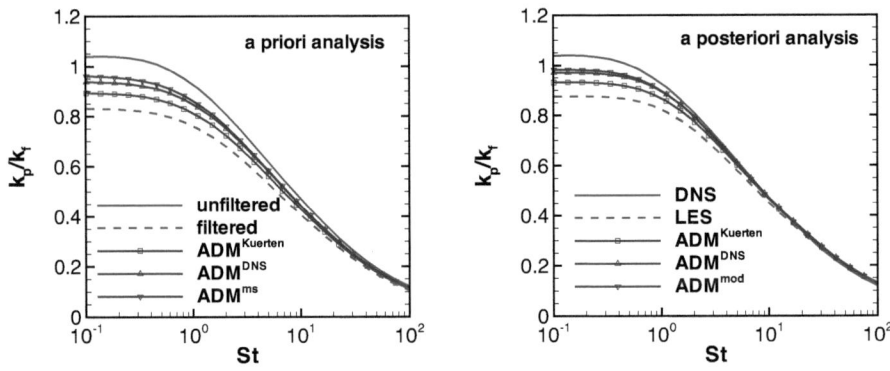

Figure 6.9: A priori (left) and a posteriori (right) analysis of ADM, particle kinetic energy (second moment in particle velocity).

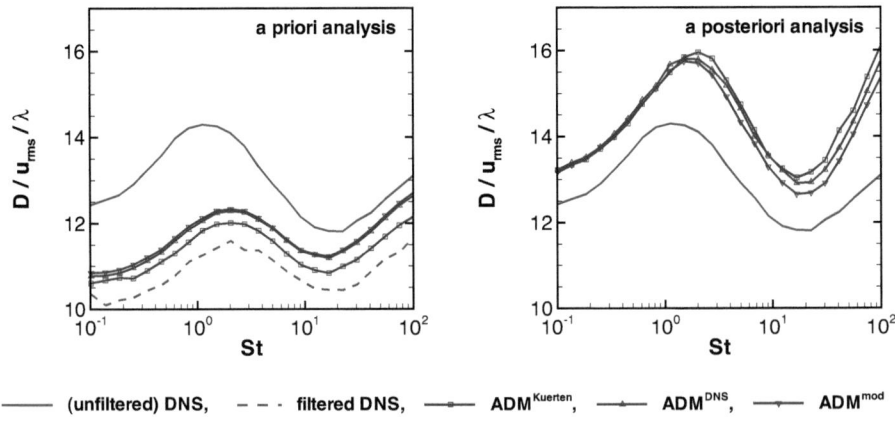

Figure 6.10: A priori (left) and a posteriori (right) analysis of ADM, rate of dispersion $D = 4k_p t_p$ (second moment in particle position). Result from LES without particle-LES model (not shown for reasons of clarity) is very similar to the results from LES with ADM.

… estimated by ADM. Particle dispersion is not predicted correctly by ADM, independent of Stokes number. Analytical computations and a priori analysis showed that ADM underestimates the rate of dispersion. On the other hand, the a posteriori analysis showed that ADM overestimates dispersion due to an approximation error in the fluid-LES model. Thus, with ADM the second moment in particle velocity is predicted correctly for high Stokes number but the second moment in particle position is error prone, independent of Stokes number.

6.4.3 Numerical assessment of the Langevin-based models

For the stochastic models, the a priori analysis was conducted differently than for ADM. For ADM, the particles were traced along the path computed from DNS. For the stochastic models, this would be in contradiction to the model assumptions. Both stochastic models take explicitly into account that the particle path depends on the small scale fluctuations which are modelled (see Shotorban & Mashayek, 2005; Fede et al., 2006). Therefore here the particle paths were computed from the modelled fluid velocity.

Concerning T_L and Γ, the analytical assessment showed that these model parameters should be set to $T_L = -\frac{1}{\Gamma_{ij}} = \frac{4k_{sgs}}{3C_0\epsilon}$ in forced isotropic turbulence, cf. equation (6.61). On the other hand, Shotorban & Mashayek (2005) and Fede et al. (2006) recommend $T_L = -\frac{1}{\Gamma_{ij}} = \frac{k_{sgs}}{(1/2+3/4C_0)\epsilon}$, independent of the flow configuration. Therefore both approaches were tested in the a priori analysis. The results from the a priori analysis are not very promising. Therefore the a posteriori analysis was only conducted with one choice for T_L, namely $T_L = -\frac{1}{\Gamma_{ij}} = \frac{k_{sgs}}{(1/2+3/4C_0)\epsilon}$.

For the a priori analysis, the model of Shotorban & Mashayek (2005) is closed with this choice of parameters. For the model of Simonin et al. (1993), the subgrid stress tensor τ must be computed additionally. Here, two approaches are possible. Either one computes τ from the original definition, equation (3.5), or one takes into account that the underlying fluid-LES model is based on an eddy viscosity hypothesis, equation (3.6). The first approach would involve less modelling assumptions. Therefore this approach would simplify comparison between results from filtered and unfiltered DNS. The second approach simplifies comparison of results from filtered DNS and LES. Therefore this approach was chosen for the present work.

For LES both Langevin-based models need additional estimates for k_{sgs} and ϵ. Here, the present work follows the recommendations of the original authors. Based on the work of Lilly (1967), the subgrid kinetic energy is estimated by

$$k_{sgs} = \frac{\nu_t^2}{C_2^2 \Delta^2}. \tag{6.72}$$

Δ denotes the LES cell size.

The constant C_2 is a model constant. Lilly (1967) proposed $C_2 = 0.094$, Yoshizawa (1982) proposed $C_2 = 0.065$. It is clear that the best value for C_2 depends on the configuration. For the present simulations, C_2 was set such that for $St = 0$ the kinetic energy of the particles approximates the DNS result reasonably well. This gives $C_2 = 0.033$.

Following Yoshizawa (1982), ϵ can be estimated by

$$\epsilon = C_\epsilon \frac{k^{3/2}}{\Delta}. \tag{6.73}$$

Following the recommendation of Gicquel *et al.* (2002) and Berrouk *et al.* (2007), C_ϵ was set to 1. C_0 was set to $C_0 = 2.1$.

Concerning the effect of variations of the model parameters, the reader is referred to Berrouk *et al.* (2007). The present chapter focuses on the structure of the models and not on the parameters.

The stochastic differential equations (6.5) and (6.8) were solved by an Euler-Maruyama scheme (see e.g. Kloeden & Platen, 2000). The stiff terms $-u_{f@p,i}^{Sho}/T_L$ and $\Gamma_{ij}u_{f@p,j}^{Sim'}$ were discretised implicitly. Shotorban & Mashayek (2005) and Fede *et al.* (2006) used an explicit Euler-Maruyama scheme. These authors focused on small Stokes numbers. In the present simulations no significant differences between explicit and implicit discretisation was found at small Stokes numbers. At high Stokes numbers, the explicit scheme was found to produce significantly worse results. In particular, the kinetic energy seen by the particles explodes at high Stokes numbers when using an explicit scheme. It should be noted that the terms under consideration are linear and therefore implicit schemes do not produce any computational overhead.

In the following, 'Sho' denotes results from the model proposed by Shotorban & Mashayek (2005) and 'Sim' denotes results for the model proposed by Simonin *et al.* (1993). 'filtered' refers to statistics of the filtered fluid velocity. In contrast to ADM, in the 'filtered' simulations, the particle trajectories were computed from the filtered fluid velocity and not from the unfiltered velocity. The reasons for this were stated above.

Second moments computed with the Langevin-based models

The testcase for the Langevin-based models is again isotropic turbulence at $Re_\lambda = 52$. Figure 6.11 shows the kinetic energy of the fluid seen by the particles. In particular the results from the a priori analysis are very disappointing. Up to $St \approx 5$, the prediction of the kinetic energy seen by the particles from the Langevin-based models is not very well but still within the range of the DNS result. Around $St = 5$, the model proposed by Simonin *et al.* (1993) leads to an excessive overshoot in the kinetic energy seen by the particles. This is because an error in the modelled quantity $\mathbf{u}_{f@p}^{Sim}$ further increases δ, the error of the kinetic energy seen by the particles, cf. equations (6.42) and (6.57). This means that with the model of Simonin *et al.* (1993), errors build up. It is interesting to denote that the authors of both models only published results up to $St = 5$.

As expected, the choice $T_L = -\frac{1}{\Gamma_{ij}} = \frac{k_{sgs}}{(1/2+3/4C_0)\epsilon}$ leads to smaller kinetic energy than $T_L = -\frac{1}{\Gamma_{ij}} = \frac{4k_{sgs}}{3C_0\epsilon}$, cf. equations (6.59) and (6.63). However, both choices lead to poor performance in the a priori analysis.

Surprisingly, the result from the a posteriori analysis is very much better. Here, both models predict the kinetic energy seen by the particles relatively well. One reason is that for

6 Presentation and assessment of existing particle-LES models

the a posteriori analysis the model parameter C_2 was set such that the kinetic energy seen by inertia free particles is close to the DNS result. Another reason is linked to the fact that the velocity gradients of the filtered field are about 5% higher than those of the LES field,

$$\left\|\frac{\partial \langle \mathbf{u}_f \rangle}{\partial x_i}\right\| \approx 1.05 \left\|\frac{\partial [\mathbf{u}_f]}{\partial x_i}\right\|. \tag{6.74}$$

As in chapter 5, $\langle \cdot \rangle$ denotes filtered DNS field and $[\cdot]$ denotes LES field. The velocity gradient affects the error δ of the kinetic energy seen by the particles, cf. equations (6.41) and (6.57). Therefore the higher gradients lead to larger errors and in particular for the model of Simonin et al. (1993), the errors build up as mentioned above.

The error in the kinetic energy of the particles is simply a consequence of the error in the kinetic energy seen by the particles, cf. figure 6.12. The a priori analysis is very interesting for the model of Simonin et al. (1993) with $-\frac{1}{\Gamma_{ij}} = \frac{4k_{sgs}}{3C_0\epsilon}$. At low Stokes numbers, this model predicts particle kinetic energy well but at high Stokes numbers, this model overpredicts kinetic energy. The same can be observed in the a posteriori analysis for the model of Shotorban & Mashayek (2005) with $T_L = \frac{k_{sgs}}{(1/2+3/4C_0)\epsilon}$.

The same observations can be made in the rate of dispersion, figure 6.13. The result from the a priori analysis is very discouraging, the a posteriori results are somewhat better concerning accurracy of the models. Nevertheless, these tests show that both models do not neccesarily improve the result of the LES in comparison to an LES without particle-LES model or LES with ADM. In particular at high Stokes numbers it might be recommendable to use no model instead of one of the stochastic models, in accordance with the numerical results of Shotorban & Mashayek (2005).

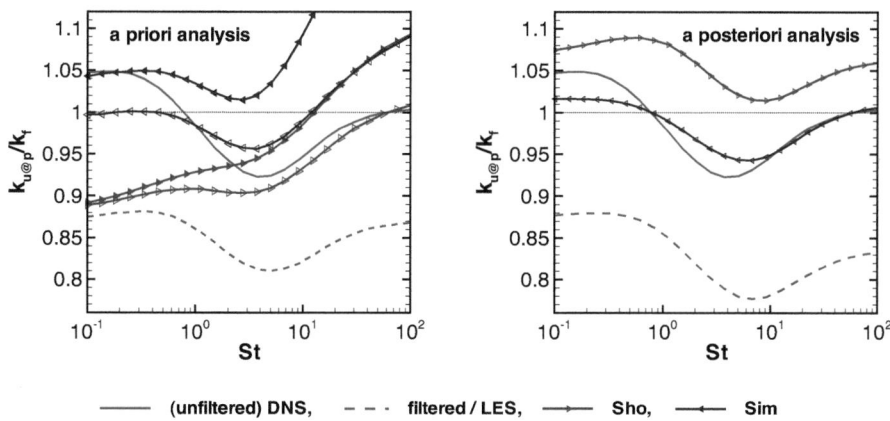

Figure 6.11: A priori (left) and a posteriori (right) analysis of the Langevin-based models, kinetic energy seen by particles (second moment in fluid velocity seen by particles). Open symbols: $T_L = -\frac{1}{\Gamma_{ij}} = \frac{k_{sgs}}{(1/2+3/4C_0)\epsilon}$ following the original papers, cf. (6.6). Filled symbols: $T_L = -\frac{1}{\Gamma_{ij}} = \frac{4k_{sgs}}{3C_0\epsilon}$ following the results from the analytical assessment, cf. equation (6.61).

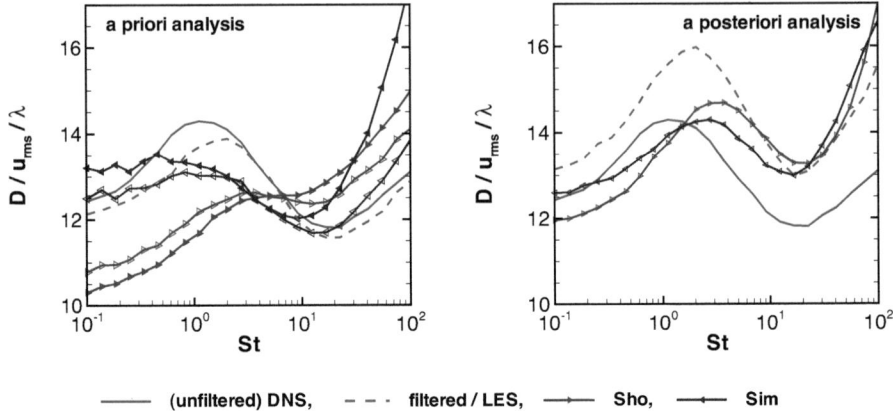

Figure 6.13: A priori (left) and a posteriori (right) analysis of the Langevin-based models, rate of dispersion (second moment in particle position). Open symbols: $T_L = -\frac{1}{\Gamma_{ij}} = \frac{k_{sgs}}{(1/2+3/4C_0)\epsilon}$ following the original papers, cf. (6.6). Filled symbols: $T_L = -\frac{1}{\Gamma_{ij}} = \frac{4k_{sgs}}{3C_0\epsilon}$ following the results from the analytical assessment, cf. equation (6.61).

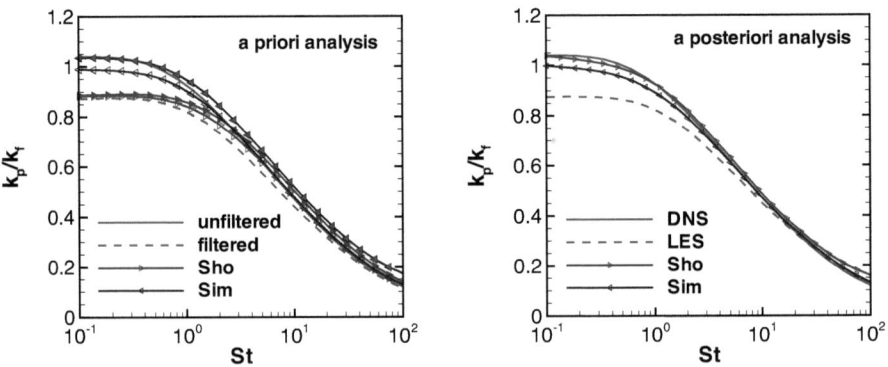

Figure 6.12: A priori (left) and a posteriori (right) analysis of the Langevin-based models, particle kinetic energy (second moment in particle velocity). Open symbols: $T_L = -\frac{1}{\Gamma_{ij}} = \frac{k_{sgs}}{(1/2+3/4C_0)\epsilon}$ following the original papers, cf. (6.6). Filled symbols: $T_L = -\frac{1}{\Gamma_{ij}} = \frac{4k_{sgs}}{3C_0\epsilon}$ following the results from the analytical assessment, cf. equation (6.61).

6 Presentation and assessment of existing particle-LES models

6.4.4 Preferential concentration

Preferential concentration was analysed by numerical means only. Figure 6.14 shows the result of the a posteriori analysis. An a priori analysis was not conducted for preferential concentration because of the high computational requirements (see Gobert & Manhart, 2009). Evidently the results from the stochastic models are far off the other results. This means that the stochastic terms destroy preferential concentration.

On the other hand, the result from ADM is quite satisfactory. Although ADM slightly overestimates preferential concentration for Stokes numbers smaller than one, ADM performs much better than the stochastic models. In particular, maximal preferential concentration is attained around $St = 1$ with ADM but not with the stochastic models. According to this result, ADM is advantageous over the stochastic models concerning preferential concentration.

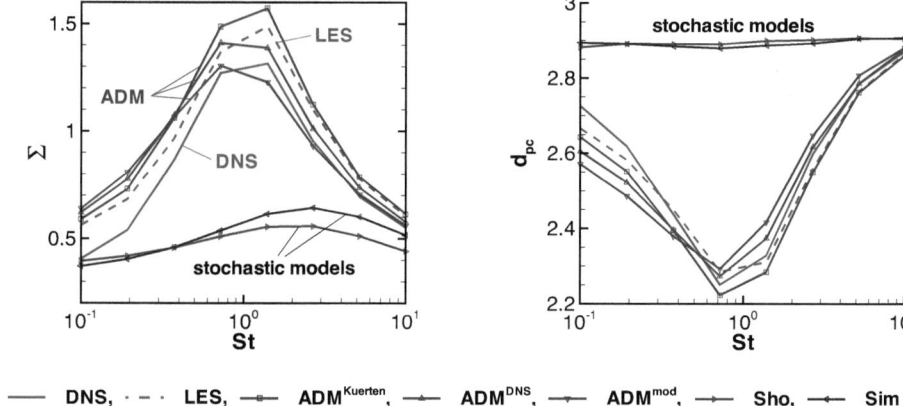

Figure 6.14: A posteriori analysis of preferential concentration. For the stochastic models (i.e. the models proposed by Shotorban & Mashayek (2006) and Simonin et al. (1993)), T_L and Γ were set to $T_L = -\frac{1}{\Gamma_{ij}} = \frac{k_{sgs}}{(1/2+3/4 C_0)\epsilon}$, following the original papers, cf. (6.6).

6.5 Conclusions of chapter 6

For Large Eddy Simulation of particle-laden flow, there exists a multitude of models for the effect of the unresolved scales on the particles. The present chapter focuses on the selection of such a model prior to simulation. To this end, an analytical method for model assessment is presented and applied on three very prominent models, namely approximate deconvolution (ADM) as proposed by Kuerten (2006b) and two stochastic models, based on the works of Shotorban & Mashayek (2006) and Simonin et al. (1993).

The analysis consists of two parts, namely analytical computations and numerical simulation. The analytical part is based on assessment of the statistical moments A1 to A6 from chapter 4. For the analytical computations, preferential concentration was neglected, Stokes drag was assumed to be linear and all model parameters were assumed to be set optimal. In particular the latter assumption means that the analysis does not focus on model constants but on the structure of the models.

The first moments are average particle position and particle velocity. According to the analysis presented, these moments are error prone for all models considered unless the flow's properties (such as homogeneity) cancel out these errors.

The error magnitude depends on Stokes number, flow structure and LES resolution. For ADM, only scales which are smaller than the LES grid contribute to the error. For the stochastic models, in the present work a wave- and Stokes number dependent function is defined which characterises the influence of eddies of a specific size on the error in the first moments. The scalar product of this function with the energy spectrum of the flow is an estimate for the first moment error.

According to the analysis presented, at low Stokes number the stochastic models show less error in the first moments than ADM whereas for high Stokes number ADM performs better. The critical Stokes number depends on the resolution of the LES grid and the energy spectrum of the flow. The reader is reminded that this result was obtained assuming optimal parameters for all models. This means that the defects cannot be remedied by a better choice of model constants.

Furthermore, the error in the stochastic models was traced back to a defect in the convective term. This defect is greater in the model proposed by Shotorban & Mashayek than in the model proposed by Simonin *et al.*, promising a higher accuracy for Simonin *et al.*'s model concerning first moments.

The second moments are average kinetic energy and turbulent dispersion of the particles. These moments are only of concern if the defects from the first moments are negligible. For the stochastic models this means that Stokes number must be small. It was shown that, if Stokes number is small enough, then the stochastic models predict first and second moments correctly. Again, this result presumes optimal parameters. Furthermore it was shown that ADM leads to defects in the second moments for any Stokes number.

It should be noted that the presented analytical computations are valid for arbitrary flows unless noted otherwise. In specific configurations the models might perform very much better than predicted herein. In particular, all first moments in isotropic turbulence are zero, predicted correctly by all models.

The analysis shows that it is very difficult to predict the first moments exactly for arbitrary configurations but that with the correct choice of LES model(s), the defects can be minimised.

The second part of the present chapter, the analysis by numerical simulation, considers the same models as the analytical part. In contrast to the analytical computation, Stokes drag is not assumed to be linear but the non-linear particle transport equation (2.32) was solved instead. Another difference to the analytical computations is that in numerical simulation preferential concentration occurs. The testcase was isotropic turbulence at $Re_\lambda = 52$. Evidently first moments are not of concern in this flow.

The analysis was conducted by a priori and a posteriori analysis. Both analyses serve for model assessment and the a priori analysis serves additionally for validation of the analytical computations.

ADM was implemented in three different ways, one following Kuerten (2006b) and two

6 Presentation and assessment of existing particle-LES models

approaches via optimisation against spectra. It should be noted that 'ADM by optimisation against spectra' does not refer to ADM as proposed by Hickel et al. (2006). They presumed a flow at infinite Reynolds number for computing the SGS viscosity ν_t whereas the present work uses model spectra for computing the ADM coefficients.

With all three approaches the analytical predictions could be verified. In particular, the defect in the rate of dispersion could be observed. In accordance with the analytical computations, ADM was found to underestimate the rate of dispersion if applied on the filtered DNS field. In LES however, ADM was found to overpredict the rate of dispersion. This discrepancy was explained as an effect of approximation errors in the fluid-LES model.

The stochastic models showed very poor performance in the numerical simulations. In particular for high Stokes numbers, LES or filtered DNS with stochastic models showed larger difference to DNS results than LES or filtered DNS without particle-LES model.

Finally, the models were analysed with repect to preferential concentration. Here, only results from numerical simulation are presented but not from analytical computations. The reader is reminded that chapter 5 showed already that no particle-LES model is necessary to predict preferential concentration correctly for the present testcases.

ADM was found to modify preferential concentration slightly. It seems that for $St < 1$, ADM tends to overpredict preferential concentration whereas for $St > 1$, ADM has a slighty negative effect on preferential concentration, i.e., LES without particle-LES model is slightly closer to DNS than LES with ADM. On the other hand, the stochastic models were found to destroy preferential concentration. With these models, preferential concentration is merely observable.

According to these results, the stochastic models are not recommendable because, in dependence of the configuration, LES without particle-LES model can perform better, sometimes even tremendously better, than LES with a stochastic particle-LES model. On the other hand, results from ADM are quite promising. At least ADM was found to lead to an improvement in first and second moments for all Stokes numbers. Preferential concentration was slightly mispredicted by ADM.

On the other hand, ADM only enhances the resolved scales but does not actually model scales which cannot be represented on the grid. At high Reynolds numbers, where the LES grid is very coarse due to computational limitations, ADM can be expected to perform worse than at low Reynolds numbers. For LES at high Reynolds numbers a new particle-LES model is needed. Such a model is presented in the following chapter.

7 A Novel Particle-LES Model based on Spectrally Optimised Interpolation (SOI)

> *Essentially, all models are wrong, but some models are useful.*
> George E. P. Box

In this chapter, a new model for the effect of unresolved scales on particles is presented. The model can be regarded as an extension of ADM towards higher wavenumbers. It is constructed such that the interpolation of fluid velocity on the particle positions reconstructs unresolved scales. The model takes advantage of the interpolation error such that the spectrum seen by the particles attains a model spectrum. Thus, the model is called Spectrally Optimised Interpolation (SOI).

In this chapter, the SOI model is only presented in the framework of homogeneous isotropic turbulence. Possible extensions for more general configurations are only briefly sketched.

As mentioned above, the model is based on the spectral interpretation of interpolation errors and is an extension of the idea behind ADM. These two topics are discussed in detail in section 7.1. Based on these analyses, the complete model is developed in section 7.2. Then, in sections 7.3 and 7.4, the model is discussed and assessed by analytical computation and numerical experiment. A comparison of the new model against other particle-LES models can be found in section 7.5. Finally, section 7.5 contains ideas for extensions of the model regarding non-homogeneous turbulence.

In the following, all considerations are presented for an equidistant grid with $\Delta x = \Delta y = \Delta z = 1$. Extensions to non-equidistant grids are mainly technical modifications. These are not very interesting and, therefore, are not presented in this work.

7.1 Preliminary considerations

The present section discusses the bases of the particle-LES model that will be developed in section 7.2.

The first basis is the spectral interpretation of interpolation errors. Therefore, section 7.1.1 considers the effect of interpolation on the spectrum seen by particles.

The second basis of the model is extending the idea behind ADM. Therefore, section 7.1.2 discusses ADM again, with a focus on the spectrum seen by the particles.

7.1.1 The spectrum seen by particles

The present section addresses the effect of interpolation on the spectrum seen by the particles. The particle-LES model, which is presented in section 7.2, takes advantage of this effect.

In chapter 6, several particle-LES models were presented. None of these models take into account the effect of interpolation. All models inherently assume that interpolation does not modify the spectrum of the flow, i.e., that the spectrum of the flow seen by the particles equals the spectrum of the flow computed from the discretised flow field. This assumption is correct for spectral interpolation, i.e., interpolation based on Fourier modes. In general, spectral interpolation is too expensive in terms of numerical costs. Instead, polynomial interpolation schemes are implemented. These modify the spectrum seen by the particles, and in particular, they introduce high wavenumber content, as explained in the present section.

For simplicity, all results are first formulated for one-dimensional interpolation and then extended to three-dimensional interpolation.

Considerations in one dimension

Consider a one-dimensional grid x_j and denote the grid points by $x_j = j$ such that the particle resides at some position $x \in [x_0, x_1[$. In particular, allow for negative values of j. Denote the fluid velocity in x_j by u_j. Consider interpolation schemes that can be formulated as a linear combination of the sample values u_j and some functions $w_j(x)$

$$u_{f@p}(x) = \sum_j w_j (x - x_j) u_j. \tag{7.1}$$

The functions w_j define the scheme. For polynomial interpolation schemes, w_j are the Lagrangian base functions,

$$w_{j,poly}(\xi) = \prod_{i \neq j} \left(\frac{\xi}{j-i} + 1 \right). \tag{7.2}$$

Taking advantage of the fact that $x - x_j \in [-j, 1-j[$, the subscript j in w_j can be omitted. More formally, construct a function w by

$$w(\xi) := \sum_j w_j(\xi) \delta_{\xi \in [-j, 1-j[}. \tag{7.3}$$

where δ denotes the Kronecker delta function. w is referred to as interpolation kernel. With this notation, the interpolation scheme reads

$$u_{f@p}(x) = \sum_j w(x - x_j) u_j. \tag{7.4}$$

With formula (7.4), the fluid velocity seen by particles $u_{f@p}$ is a continuous function and its

Fourier transform can be computed for an arbitrary wavenumber κ,

$$\mathcal{FT}(u_{f@p})(\kappa) = \int_{-\infty}^{\infty} u_{f@p}(x) e^{-i\kappa x} \, dx = \int_{-\infty}^{\infty} \sum_j w(x - x_j) u_j e^{-i\kappa x} \, dx$$

$$= \mathcal{FT}(w)(\kappa) \underbrace{\sum_j u_j e^{-i\kappa x_j}}_{=\mathcal{FT}(u_f)(\kappa)}. \quad (7.5)$$

This means that the spectrum seen by the particle equals the spectrum of the interpolation kernel, multiplied by the continuous spectrum of the sample data,

$$|\mathcal{FT}(u_{f@p})(\kappa)|^2 = |\mathcal{FT}(w)(\kappa)|^2 |\mathcal{FT}(u_f)(\kappa)|^2 \quad (7.6)$$

It should be noted that $\mathcal{FT}(u_f)(\kappa)$ is not zero beyond the cutoff wavenumber $\kappa_c = \pi$. For example for $\kappa = \kappa_c + \kappa'$, $0 < \kappa' < \kappa_c$, it holds

$$\mathcal{FT}(u_f)(\kappa) = \sum_j u_j e^{-i(2\kappa_c + (\kappa' - \kappa_c))x_j} = \sum_j u_j e^{-i2\pi j} e^{i(\kappa_c - \kappa')x_j} = (\mathcal{FT}(u_f)(\kappa_c - \kappa'))^*. \quad (7.7)$$

Here, $(\cdot)^*$ denotes the complex conjugate. Following this argument, the spectrum $|\mathcal{FT}(u_f)(\kappa)|^2$ can be computed by piecewise reflection of the spectrum for $\kappa < \kappa_c$ (cf. figure 7.1). This result is well known under the name of aliasing.

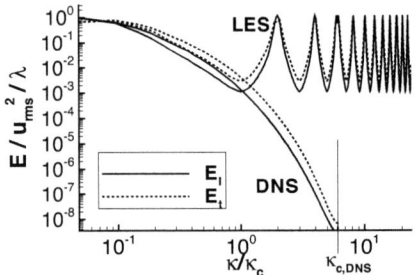

Figure 7.1: 1D-spectra of isotropic turbulence at $Re_\lambda = 52$ (cf. chapter 5). Continuous lines: longitudinal spectra, dashed lines: transverse spectra. LES spectra are prolonged for $\kappa > \kappa_c$ according to equation (7.7). Scaling on the wavenumber axis is based on LES cutoff wavenumber.

Considerations in three dimensions

The argumentation presented above is based on one-dimensional considerations. Turbulence is three-dimensional and must be treated as such. Therefore, the following focuses on issues in three-dimensional interpolation.

Three-dimensional interpolation is an issue by itself, in particular if arbitrarily located data needs to be interpolated (see e.g. Kincaid & Cheney, 2001). However, this is not the case in the present work because only data on Cartesian grids are considered. Again, all results are presented for equidistant grids $\Delta x = \Delta y = \Delta z = 1$ for reasons of simplicity, although they are also valid for non-equidistant grids.

7 A Novel Particle-LES Model based on Spectrally Optimised Interpolation (SOI)

In three dimensions, the interpolation formula (7.1) must be extended. In general, the interpolation kernel becomes a matrix $\mathbf{W}$:

$$\mathbf{u}_{f@p}(x) = \sum_{i,j,k} \mathbf{W}\left(\xi_i, \eta_j, \zeta_k\right) \mathbf{u}_f(x_i, y_j, z_k) \tag{7.8}$$

$$\text{with} \qquad \xi_i = x - x_i, \qquad \eta_j = y - y_j, \qquad \zeta_k = z - z_k. \tag{7.9}$$

The one-dimensional spectra seen by the particles read

$$\begin{aligned} E_{ij}^{f@p}(\kappa) &= \int_{\mathbf{R}^3} \delta_{k_j}(\kappa) \left|\mathcal{FT}(u_{f@p,i})(\mathbf{k})\right|^2 \, d\mathbf{k} = \\ &= \int_{\mathbf{R}^3} \delta_{k_j}(\kappa) \left|(\mathcal{FT}(W_{i,n})(\mathbf{k}))^* \, \mathcal{FT}(u_{f,n})(\mathbf{k})\right|^2 \, d\mathbf{k} \end{aligned} \tag{7.10}$$

where $\mathbf{k}$ denotes the three-dimensional wave vector and $\mathcal{FT}(\mathbf{W})(\mathbf{k})$ is the component-wise three-dimensional Fourier transform of the matrix $\mathbf{W}$.

For the one-dimensional case, the spectrum seen by the particles $|\mathcal{FT}(u_{f@p})|^2$ equals the product of $|\mathcal{FT}(w)|^2$ and $|\mathcal{FT}(u_f)|^2$ (cf. equation (7.5)). This means that, in 1D, the spectrum seen by the particles is simply the product of the transfer function of the interpolation and the spectrum of the flow. In three dimensions, this does not hold,

$$E_{ij}^{f@p}(\kappa) \neq \left(\int_{\mathbf{R}^3} \delta_{k_j}(\kappa) \left|\mathcal{FT}(W_{i,n})(\mathbf{k})\right|^2 \, d\mathbf{k} \right) E_{nj}^f(\kappa). \tag{7.11}$$

However, the reflection property (7.7) holds in three dimensions,

$$\begin{aligned} E_{ij}(\kappa) &= \int_{\mathbf{R}^3} \delta_{k_j}(\kappa) \left|\mathcal{FT}(u_{i,f})(\mathbf{k})\right|^2 \, d\mathbf{k} = \int_{\mathbf{R}^3} \delta_{k_j}(\kappa) \left|\mathcal{FT}(u_{i,f})(\mathbf{k} + (\kappa_c + \kappa' - k_j)\mathbf{e}_j)\right|^2 \, d\mathbf{k} \\ &= \int_{\mathbf{R}^3} \delta_{k_j}(\kappa_c - \kappa') \left|\mathcal{FT}(u_{i,f})(\mathbf{k})\right|^2 \, d\mathbf{k} = E_{ij}(\kappa_c - \kappa') \end{aligned} \tag{7.12}$$

with the j-th unit vector $\mathbf{e}_j$ and $\kappa = \kappa_c + \kappa'$, $0 < \kappa' < \kappa_c$. For the model proposed in section 7.2, this property plays a crucial role.

The reflection property holds for the one-dimensional spectra E_{ij} but not for the energy spectrum function E. Therefore, the present section focuses on one-dimensional spectra instead of the energy spectrum function.

Spectra from standard 3D interpolation schemes

To simulate particle-laden flow, commonly used interpolation schemes are semi-linear, tri-linear and fourth-order Lagrangian interpolation. In the following, these schemes are analysed with respect to the spectrum seen by the particles.

Semi-linear interpolation means that each velocity component is interpolated linearly in its respective direction. For the other directions, nearest-neighbour interpolation is conducted. This means that $u_{f,i}$ is linear in x_i and piecewise constant in $x_j, j \neq i$. This interpolation scheme is the simplest conservative interpolation scheme for staggered grids. The corresponding interpolation kernel $\mathbf{W}_{sl}$ reads

$$\mathbf{W}_{sl}(\xi, \eta, \zeta) = \begin{pmatrix} w_{sl}(\xi, \eta, \zeta) & 0 & 0 \\ 0 & w_{sl}(\eta, \xi, \zeta) & 0 \\ 0 & 0 & w_{sl}(\zeta, \xi, \eta) \end{pmatrix} \quad (7.13)$$

with

$$w_{sl}(\xi, \eta, \zeta) = w_{lin}(\xi) \, w_{const}(\eta) \, w_{const}(\zeta). \quad (7.14)$$

w_{lin} denotes the longitudinal interpolation kernel which is defined by

$$w_{lin}(\xi) = \begin{cases} 1 - |\xi| & \text{for } |\xi| < 1 \\ 0 & \text{otherwise} \end{cases} \quad (7.15)$$

and w_{const} denotes the transverse interpolation kernel, defined by

$$w_{const}(\eta) = \begin{cases} 1 & \text{for } |\eta| < \frac{1}{2} \\ 0 & \text{otherwise} \end{cases}. \quad (7.16)$$

w_{const} is simply a rectangular function. w_{lin} is shown in figure 7.2.

Semi-linear interpolation is often referred to as 'second-order interpolation', although the interpolation of each component $u_{f,i}$ is only second-order in the direction of x_i but not in $x_j, j \neq i$. If one interpolates the velocity components linearly in all directions, then one obtains trilinear interpolation. Trilinear interpolation is second-order in all directions but not conservative (cf. Meyer & Jenny, 2004). The kernel for trilinear interpolation reads

$$\mathbf{W}_{tl}(\xi, \eta, \zeta) = w_{lin}(\xi) \, w_{lin}(\eta) \, w_{lin}(\zeta) \, \mathbf{I}. \quad (7.17)$$

A straightforward extension of trilinear interpolation to a higher order is Lagrangian fourth-order interpolation. Its kernel reads

$$\mathbf{W}_{cub}(\xi, \eta, \zeta) = w_{cub}(\xi) \, w_{cub}(\eta) \, w_{cub}(\zeta) \, \mathbf{I} \quad (7.18)$$

where $\mathbf{I}$ denotes identity and

$$w_{cub}(\xi) = \begin{cases} (-\xi + 1)(-\xi/2 + 1)(-\xi/3 + 1) & \text{for } \xi \in [1, 2[\\ (\xi + 1)(-\xi + 1)(-\xi/2 + 1) & \text{for } \xi \in [0, 1[\\ (\xi/2 + 1)(\xi + 1)(-\xi + 1) & \text{for } \xi \in [-1, 0[\\ (\xi/3 + 1)(\xi/2 + 1)(\xi + 1) & \text{for } \xi \in [-2, -1[\\ 0 & \text{otherwise} \end{cases} \quad (7.19)$$

w_{cub} is also shown in figure 7.2. w can be interpreted as the answer of the interpolation scheme to a Dirac peak, i.e., to $u_0 = 1$ and $u_i = 0$ for $i \neq 0$. This explains the undershoots of the fourth-order scheme in $|\xi| \in]1, 2[$.

7 A Novel Particle-LES Model based on Spectrally Optimised Interpolation (SOI)

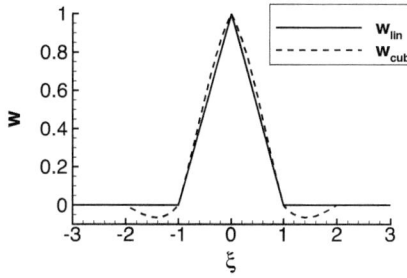

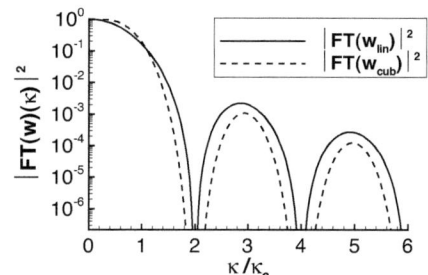

Figure 7.2: Interpolation kernels for piecewise linear and cubic interpolation.

Figure 7.3: Transfer functions for piecewise linear and cubic interpolation.

Figure 7.3 shows the transfer functions $|\mathcal{FT}(w_{lin})|^2$ and $|\mathcal{FT}(w_{cub})|^2$. It should be noted that the spectrum seen by the particles is not equal to the product of the spectrum of the flow and these transfer functions (cf. equation (7.11)). Nevertheless, the transfer functions show some important properties of the interpolation schemes. Both schemes lead to significant damping close to the cutoff wavenumber (note the logarithmic scale). Furthermore, both schemes lead to periodic enhancement at unresolved scales. For $\kappa < \kappa_c$, the fourth-order scheme leads to less damping than the second-order scheme, whereas for $\kappa > \kappa_c$, the higher order scheme shows stronger damping.

The last scheme analysed here is a linear conservative interpolation scheme proposed by Gobert et al. (2006). Its kernel reads

$$\mathbf{W}_{cl}(\xi,\eta,\zeta) = \begin{pmatrix} w_{cl}(\xi,\eta,\zeta) & 0 & 0 \\ 0 & w_{cl}(\eta,\xi,\zeta) & 0 \\ 0 & 0 & w_{cl}(\zeta,\xi,\eta) \end{pmatrix} \quad (7.20)$$

with

$$w_{cl}(\xi,\eta,\zeta) = w_{cl}^l(\xi,\eta,\zeta)\, w_{cl}^t(\eta)\, w_{cl}^t(\zeta). \quad (7.21)$$

The longitudinal and transverse interpolation kernels w_{cl}^l and w_{cl}^t are

$$w_{cl}^l(\xi,\eta,\zeta) = \begin{cases} 1 - |\xi| & \text{for } |\xi| < 1 \text{ and } \max\{|\eta|,|\zeta|\} < 0.5 \\ 0.5 & \text{for } |\xi| < 1 \text{ and } \max\{|\eta|,|\zeta|\} > 0.5 \\ 0 & \text{otherwise} \end{cases} \quad (7.22a)$$

$$w_{cl}^t(\xi) = \begin{cases} 1 & \text{for } |\xi| < 1/2 \\ \frac{1}{2} - \frac{|\xi|}{2} & \text{for } 1/2 < |\xi| < 3/2 \\ 0 & \text{otherwise} \end{cases} \quad (7.22b)$$

This scheme was used throughout the present work to quantify interpolation errors. Results from fourth-order interpolation were compared against results from this interpolation scheme.

It should be noted that $\mathbf{W}_{sl}$ and $\mathbf{W}_{cl}$ lead to discontinuities in $\mathbf{u}_{f@p}$. This has consequences for the respective transfer functions. If a function w is n-times differentiable, then the absolute value of its Fourier transform $|\mathcal{FT}(w)(\kappa)|$ decays for $\kappa \to \infty$ faster than κ^n (see e.g. Strichartz, 1994). Therefore, at high wavenumbers, the Fourier transforms $\mathcal{FT}(\mathbf{W}_{tl})$ and $\mathcal{FT}(\mathbf{W}_{cub})$ can be expected to decay faster than $\mathcal{FT}(\mathbf{W}_{sl})$ and $\mathcal{FT}(\mathbf{W}_{cl})$.

This can be readily verified by computing the 1D-spectra seen by the particles. They are shown in figures 7.4 and 7.5 for all considered schemes. Again, the test case is isotropic turbulence at $Re_\lambda = 52$. For this test case, 100 million particles were regularly positioned on 100 planes, each normal to the z-axis. Then, the x-component of the interpolated fluid velocity at the particle positions was computed for a single time step. More precisely, an instantaneous value of $u_{1,f@p}(\mathbf{x})$ was computed for $\mathbf{x} = (i\widehat{\Delta x}, j\widehat{\Delta y}, k\widehat{\Delta z})$ with $i,j = 1,\ldots,1000, k = 1,\ldots,100$. $\widehat{\Delta x}, \widehat{\Delta y}$ and $\widehat{\Delta z}$ were set such that the complete computational box is covered, i.e., $\widehat{\Delta x} = \widehat{\Delta y} = L/1000$ and $\widehat{\Delta z} = L/100$. L is the length of the computational box. The data $u_{1,f@p}(\mathbf{x})$ allow the 1D spectra E_{11} and E_{12} to be computed.

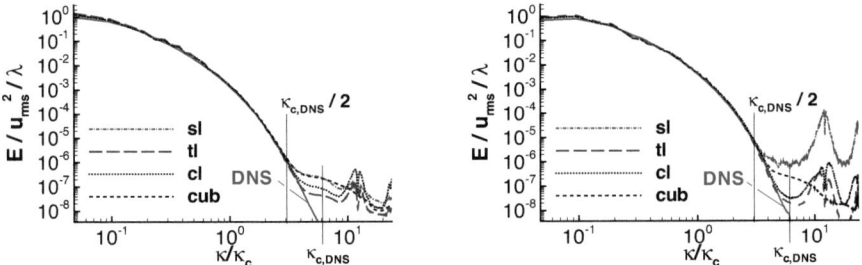

Figure 7.4: Longitudinal (left) and transverse (right) spectra seen by particles in isotropic turbulence at $Re_\lambda = 52$, computed by DNS. Dash-dotted line (sl): semi-linear interpolation, long-dashed line (tl): trilinear interpolation, dotted line (cl): second-order conservative interpolation, short-dashed line (cub): fourth-order interpolation. For reference, the spectrum computed from the grid points is also shown (continuous line).

The DNS results (figure 7.4) already show an interesting behaviour. In the range $\kappa \lesssim \kappa_{c,DNS}/2$, all interpolation schemes give identical spectra. However, starting around half the DNS cutoff wavenumber, even the fourth-order scheme shows significant deviations from the spectrum computed from the flow. This behaviour was not expected from the transfer functions shown in figure 7.4. However, this is simply a consequence of the difference between 1D and 3D transfer functions (cf. (7.11)).

For $\kappa > \kappa_{c,DNS}$, the semi-linear, trilinear and second-order conservative scheme show the damped reflected spectrum nicely. Damping is weaker in the transverse spectrum than in the longitudinal spectrum because of the discontinuities of the schemes in the transverse direction. Accordingly, the fourth-order scheme does not show the reflections due to its smooth solution, i.e., strong damping for $\kappa > \kappa_{c,DNS}$.

The trilinear scheme shows stronger damping around κ_c than the fourth-order scheme. Therefore, the spectrum from trilinear interpolation is closer to the spectrum computed from

7 A Novel Particle-LES Model based on Spectrally Optimised Interpolation (SOI)

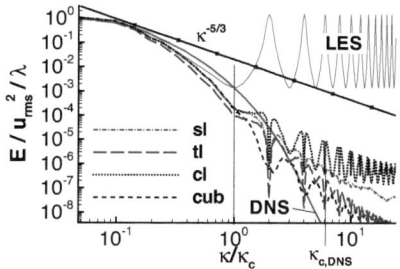 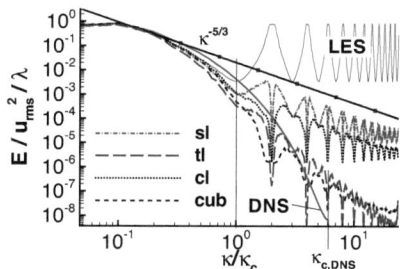

Figure 7.5: Longitudinal (left) and transverse (right) spectra seen by particles in isotropic turbulence at $Re_\lambda = 52$, computed by LES. Dash-dotted line (sl): semi-linear interpolation, long-dashed line (tl): trilinear interpolation, dotted line (cl): second-order conservative interpolation, short-dashed line (cub): fourth-order interpolation. For reference, the spectra computed from the grid points in DNS and LES (plus reflections) are also shown (continuous lines).

---

the flow than the spectrum from fourth-order interpolation.

Even more interesting are the results from LES (figure 7.5). Below the cutoff wavenumber κ_c, the spectra from semi-linear, conservative second-order and fourth-order interpolation are basically on top of each other. The trilinear scheme shows the strongest damping in that range. The 1D spectra demonstrated that it is expected that the trilinear scheme shows stronger damping in that range than the fourth-order scheme. The fact that the trilinear scheme also shows stronger damping than the semi-linear and the conservative second-order schemes is a result of the discontinuities of the latter schemes.

Beyond the cutoff wavenumber, the damped reflected spectra can be clearly identified. Because of the low Reynolds number, the DNS spectrum decays rapidly. Figure 7.5 shows that this is not the case for the spectra seen by the particles. The high order scheme shows the strongest damping in that range and, therefore, the closest similarity to DNS at that Reynolds number.

At higher Reynolds numbers, the DNS spectrum approaches the $\kappa^{-5/3}$ line. The results of the low order schemes seem to follow this trend in contrast to the fourth-order results. This can be explained because of the link between the continuity of a function and its Fourier transform. The higher order scheme produces a smoother interpolated velocity. The smoother the signal, the faster its Fourier transform decays (see e.g. Strichartz, 1994).

7.1.2 ADM revisited - ADM and interpolation

The previous section discussed the effect of interpolation on the spectrum seen by the particles with a focus on DNS and LES without a model. The present section analyses the effect of interpolation for LES with approximate deconvolution (ADM, cf. section 6.2.2). This analysis opens the possibility of extending the idea behind ADM towards higher wavenumbers, which is performed in section 7.2.

Details on the implementation of ADM were presented in section 6.4.2. In that section, three approaches for ADM were presented. One approach follows Kuerten (2006b), and the other two approaches are based on reconstruction of target spectra. The latter two approaches showed higher accuracy in section 6.4.2. Therefore, in the following, only these two approaches are considered.

The spectrum seen by particles when using ADM depends on the ADM stencil and on the interpolation scheme. The first dependency can be analysed by translating the ADM velocity field in spectral space (cf. equation (6.69a))

$$\mathcal{FT}(\mathbf{u}_f^{ADM})(\mathbf{k}) = \left(\mathcal{FT}\left(\mathbf{w}^{ADM,3D}\right)(\mathbf{k})\right)^* \mathcal{FT}(\langle \mathbf{u}_f \rangle)(\mathbf{k}). \tag{7.23}$$

Again, $(\,\cdot\,)^*$ denotes the complex conjugate. Including the effect of interpolation, one obtains (cf. equation (7.10))

$$\mathcal{FT}(\mathbf{u}_{f@p}^{ADM})(\mathbf{k}) = \left(\mathcal{FT}(\mathbf{W})(\mathbf{k})\,\mathcal{FT}\left(\mathbf{w}^{ADM,3D}\right)(\mathbf{k})\right)^* \mathcal{FT}(\langle \mathbf{u}_f \rangle)(\mathbf{k}) \tag{7.24}$$

where $\mathbf{W}$ denotes the interpolation kernel. Accordingly, the spectrum seen by particles in ADM is

$$E_{ij}^{f@p,ADM}(\kappa) = \int_{\mathbf{R}^3} \delta_{k_j}(\kappa) \left|\left(\mathcal{FT}(W_{i,\cdot})(\mathbf{k})\,\mathcal{FT}(\mathbf{w}^{ADM,3D})(\mathbf{k})\right)^* \langle \mathcal{FT}(\mathbf{u}_f)(\mathbf{k})\rangle\right|^2 \mathrm{d}\mathbf{k} \tag{7.25}$$

In order to separate the effect of interpolation from the effect of ADM, it is tempting to write $E_{ij}^{f@p,ADM}(\kappa)$ in terms of

$$\int_{\mathbf{R}^3} \delta_{k_j}(\kappa) \left|(\mathcal{FT}(W_{i,\cdot})(\mathbf{k}))^* \langle \mathcal{FT}(\mathbf{u}_f)(\mathbf{k})\rangle\right|^2 \mathrm{d}\mathbf{k} \quad \text{and} \quad \int_{\mathbf{R}^3} \left|\mathcal{FT}(\mathbf{w}^{ADM,3D})(\mathbf{k})\right|^2 \mathrm{d}\mathbf{k} \tag{7.26}$$

but this is not possible (cf. equation (7.11)). Therefore, the effect of ADM and interpolation cannot be analysed separately. Thus, an analysis of the combined effect follows.

ADM based on optimisation against DNS spectrum

First, the effect of ADM^{DNS} on the spectrum seen by particles is analysed; i.e., the ADM stencil is optimised against the DNS spectrum. Deviations between spectra from ADM and DNS are due to inherent limitations of the model and not a result of any secondary closure assumptions.

In figure 7.6, one-dimensional spectra from LES of isotropic turbulence at $Re_\lambda = 52$ are shown with and without ADM. In both simulations, the Lagrangian fourth-order interpolation $\mathbf{W}_{cub}$ was implemented. As expected, ADM enhances the longitudinal and the transverse spectra, especially in the low wavenumber range. However, figure 7.6 shows that ADM also has an effect beyond the cutoff wavenumber. Moreover, in that wavenumber range ADM enhances the spectrum seen by the particles.

Figure 7.7 shows the spectra from ADM with second- and fourth-order interpolation. The

7 A Novel Particle-LES Model based on Spectrally Optimised Interpolation (SOI)

second-order interpolation is the conservative scheme $\mathbf{W}_{cl}$. Again, the different interpolation schemes lead to different spectra, in particular at high wavenumbers. The effect is qualitatively identical to the effect without ADM, which is discussed in section 7.1.1. It is remarkable that for a wide range of wavenumbers the spectrum from low order interpolation is closer to the DNS spectrum than the spectrum from high order interpolation. This means that the enhancement of wavenumbers beyond the cutoff spectrum by low order interpolation might be a desired effect.

The analysis of ADM presented so far was based on ADM^{DNS}. In the following, the same analysis is conducted for ADM^{mod}.

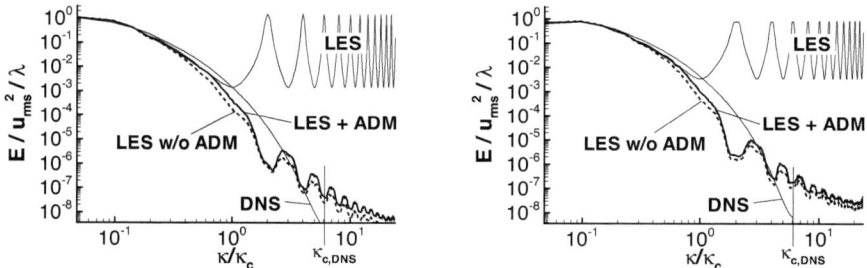

Figure 7.6: Longitudinal (left) and transverse (right) spectra seen by particles in isotropic turbulence at $Re_\lambda = 52$, computed by LES with ADM^{DNS} (continuous line) and without ADM (dashed line). For reference, the spectra computed from the grid points in DNS and LES (plus reflections) are also shown (continuous lines).

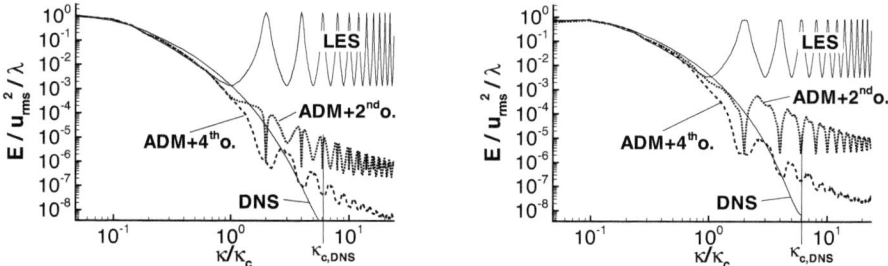

Figure 7.7: Longitudinal (left) and transverse (right) spectra seen by particles in isotropic turbulence at $Re_\lambda = 52$, computed by LES with ADM^{DNS} with fourth- and second-order interpolation. For reference, the spectra computed from the grid points in DNS and LES (plus reflections) are also shown (continuous lines).

ADM based on optimisation against model spectrum

The following focuses on ADM^{mod}, i.e., the defiltering operator that was obtained optimising against the model spectrum. Figure 6.7 from section 6.4.2 showed that this approach leads to a much stronger amplification around κ_c than the defiltering operator from the preceding section.

The one-dimensional spectra seen by the particles are depicted in figure 7.8. Again, interpolation is Lagrangian fourth-order. Here, the higher amplification factor around κ_c leads to better results at low wavenumbers. This is an effect of two errors cancelling each other. First, the fact that the model spectrum is higher than the DNS spectrum around κ_c leads to strong defiltering in that wavenumber range. Second, this effect is partially compensated by interpolation due to damping.

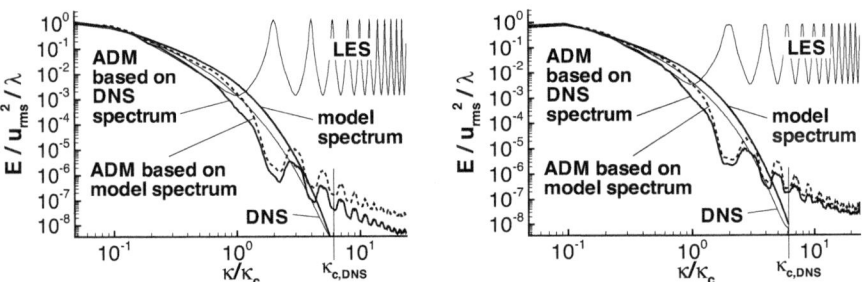

Figure 7.8: Longitudinal (left) and transverse (right) spectra seen by particles in isotropic turbulence at $Re_\lambda = 52$, computed by LES with ADM^{DNS} (continuous line) and ADM^{mod} (dashed line). For reference, the spectra computed from the grid points in DNS and LES (plus reflections) are also shown (continuous lines).

On the other hand, both ADM stencils have decay that is too rapid between κ_c and $2\kappa_c$. These observations motivate the idea of constructing an interpolation scheme that has good spectral properties in the whole spectrum. Here, 'good' refers to the difference between the spectrum seen by particles in LES and in DNS.

7.2 Construction of the SOI model

In the present section, the findings of section 7.1 are combined to construct a new particle-LES model. In 1D, the modelling strategy involves the following:

1. Compute the LES spectrum $|\mathcal{FT}(\langle u_f \rangle)(\kappa)|^2$ for $\kappa < \kappa_c$. Extend this spectrum by reflection for higher values of κ (cf. equation (7.7) and figure 7.1).

2. Define a target spectrum $E^{target}(\kappa)$ for the fluid velocity seen by the particles.

3. Search for an interpolation kernel w such that $|\mathcal{FT}(\langle u_f \rangle)|^2 |\mathcal{FT}(w)|^2 \approx E^{target}$.

7 A Novel Particle-LES Model based on Spectrally Optimised Interpolation (SOI) 123

4. Apply w in order to interpolate the fluid velocity seen by the particles.

Step 3 cannot be transferred to 3D one to one because of inequality (7.11). In the present section, a possible extension is presented.

The $\approx$-sign in step 3 is a consequence of additional admissibility conditions for the interpolation, such as compact support and order of the interpolation. Section 7.2.1 contains a list of all admissibility conditions respected. In sections 7.2.2 and 7.2.3, the model is formulated as an optimisation problem. For simplicity, the admissibility conditions are neglected in this first step. They are imposed in the second step, which is described in section 7.2.4. The complete model is summarised in section 7.2.5.

The model can be regarded as an interpolation scheme that is optimised with respect to the spectrum seen by the particles and is therefore referred to as 'Spectrally Optimised Interpolation' (SOI).

7.2.1 Properties of the model (admissibility conditions)

The idea of the novel model was laid out above. In the present and following three sections, this novel model is elaborated in detail. The present section states the properties of the model. In the following sections, a model is constructed such that these properties hold. The model is referred to as the SOI model, which stands for 'Spectrally Optimised Interpolation'.

In step 3 of the strategy presented above, replacing the $\approx$-sign with an $=$-sign, does not guarantee a compact interpolation stencil w, which leads to high computational costs. With a compact stencil, only an approximate spectral match can be attained in general; thus, the $\approx$-sign is used.

In detail, the resulting scheme has the following properties:

A1 Compact support (4-point stencil).

A2 Constant functions are interpolated exactly everywhere (first-order interpolation).

A3 Linear functions are interpolated exactly in grid points (second-order in grid points).

A4 The fluid velocity seen by the particles is continuously differentiable and twice continuously differentiable almost everywhere.

A5 Distinct interpolation in the longitudinal and transverse directions.

A6 The spectrum seen by particles is optimised with respect to a model spectrum in a least square sense (spectral accuracy).

Properties A1, A2 and A4 are commonly desired properties for accurate interpolation at low computational costs. Property A6 was already motivated above. Properties A3 and A5 require more explanation.

Property A3 is related to the fact that the resulting scheme only ensures that $u_{f@p}(x_i) = u_i$ if the data (x_i, u_i) correspond to a constant or linear function. This means that in general if a particle resides on a grid point, then it might not see the fluid velocity assigned to that grid point, likewise for ADM. Thus, the expression 'interpolation scheme' might be misleading. Monaghan (1985) refers to such schemes as smoothing schemes. However, in the present context, this term is also misleading because the scheme does not smooth but rather roughens the data. Therefore, the scheme is called 'interpolation scheme' hereafter.

Property A5 corresponds to the fact that longitudinal and transverse spectra are different. Correspondingly, it makes sense to interpolate differently in the longitudinal and transverse directions.

The properties listed above are realised by unconstrained optimisation. Properties A1 to A5 are realised by construction, and property A6 defines the target function. Properties A1 to A5 are referred to as *admissibility conditions* because they define which stencils are admissible for optimisation.

The optimisation is implemented in two pre-processing steps. In the first step, the admissibility conditions A1 to A4 are omitted for computational efficiency. The solution is a non-admissible interpolation stencil denoted by w_l^{na} and w_t^{na}. The superscript 'na' stands for non-admissible.

In the second step, another optimisation is conducted. It consists of finding admissible solutions w_l and w_t that match w_l^{na} and w_t^{na} respectively as closely as possible with respect to their transfer functions.

The first step is described in sections 7.2.2 and 7.2.3; the second step is described in section 7.2.4.

As mentioned above, the model is formulated for isotropic turbulence, and extensions to other flows are sketched in section 7.5.

7.2.2 Formulation of the model as optimisation problem

In the previous section, properties of the SOI model were listed. This section and the following two concern the implementation of these properties. They are incorporated one by one. In this section, only properties A5 and A6 are observed. The other properties are incorporated in section 7.2.4.

The model is characterised by two interpolation stencils. One stencil is for interpolation in the longitudinal direction, and the other is for interpolation in the transverse direction. They are denoted by by w_l and w_t, respectively. The three-dimensional interpolation kernel is defined by

$$\mathbf{W}_{SOI}(\xi,\eta,\zeta) = \begin{pmatrix} w_l(\xi)w_t(\eta)w_t(\zeta) & 0 & 0 \\ 0 & w_t(\xi)w_l(\eta)w_t(\zeta) & 0 \\ 0 & 0 & w_t(\xi)w_t(\eta)w_l(\zeta) \end{pmatrix}. \quad (7.27)$$

The scalar sub-kernels w_l and w_t are now specified such that the spectrum seen by the particles is optimised.

$\mathbf{W}_{SOI}$ is applied on a LES velocity field $\langle \mathbf{u}_f \rangle$. Then, the longitudinal and transverse spectra seen by the particles are

$$E_l^{SOI}(k_x) = |\mathcal{FT}(w_l)(k_x)|^2 \int_{\mathbf{R}^2} |\mathcal{FT}(w_t)(k_y)\mathcal{FT}(w_t)(k_z)\mathcal{FT}(\langle u_{f,1}\rangle)(\mathbf{k})|^2 \; dk_y \; dk_z$$

$$E_t^{SOI}(k_y) = |\mathcal{FT}(w_t)(k_y)|^2 \int_{\mathbf{R}^2} |\mathcal{FT}(w_l)(k_x)\mathcal{FT}(w_t)(k_z)\mathcal{FT}(\langle u_{f,1}\rangle)(\mathbf{k})|^2 \; dk_x \; dk_z.$$

The aim is to set w_l and w_t such that E_l^{SOI} and E_t^{SOI} approximate target spectra E_l^{target} and E_t^{target}, respectively. For simplicity, the admissibility conditions A1 to A4 from page 123 are omitted for now. Then, the optimisation problem reads

$$\text{find } w_l^{na} \text{ and } w_t^{na} \text{ such that } \left\| \begin{pmatrix} E_l^{SOI} - E_l^{target} \\ E_t^{SOI} - E_t^{target} \end{pmatrix} \right\| \to \min \qquad (7.29)$$

The Euclidean norm is used; w_l^{na} and w_t^{na} denote the solution of the optimisation problem.

7.2.3 Reduction of computational overhead: Optimisation against 1D-spectra

The previous section ended in the optimisation problem (7.29). In that form, the optimisation would be computationally expensive. Therefore in the present section the optimisation problem is reformulated such that the computational costs are reduced.

Problem (7.29) takes as input data the whole (Fourier transformed) LES field $\mathcal{FT}(\langle u_{f,1}\rangle)(\mathbf{k})$. In this formulation, for each test solution (w_l, w_t) the spectra $E_l^{SOI}(\kappa)$ and $E_t^{SOI}(\kappa)$ must be computed by two-dimensional integration. Integration must be conducted for each value of κ separately because $\mathbf{k} = (k_x, k_y, k_z)$ depends on k_x and k_y. Thus, the computational overhead for the optimisation is high.

In the present section the computational overhead is reduced by including modelling assumptions. The optimisation problem is reformulated such that $\mathcal{FT}(\langle u_{f,1}\rangle)(\mathbf{k})$ is replaced by the one-dimensional LES spectra E_l^{LES} and E_t^{LES}.

Introduce the notation

$$a = |\mathcal{FT}(w_l^{na})|^2, \qquad b = |\mathcal{FT}(w_t^{na})|^2, \qquad f = |\mathcal{FT}(\langle u_{f,1}\rangle)|^2. \qquad (7.30)$$

Then, problem (7.29) reads

$$\text{find } w_l^{na} \text{ and } w_t^{na} \text{ such that } \left\| \begin{pmatrix} E_l^{target}(x) - a(x) \int_{\mathbf{R}^2} b(y)b(z)f(x,y,z) \; dy \; dz \\ E_t^{target}(y) - b(y) \int_{\mathbf{R}^2} a(x)b(z)f(x,y,z) \; dx \; dz \end{pmatrix} \right\| \to \min. \qquad (7.31)$$

The one-dimensional spectra of the LES velocity (i.e., the spectra of the fluid, not the spectra seen by the particles) can be expressed as

$$E_l^{LES}(x) = \int_{\mathbf{R}^2} f(x,y,z) \; dy \; dz, \tag{7.32a}$$

$$E_t^{LES}(y) = \int_{\mathbf{R}^2} f(x,y,z) \; dx \; dz = \int_{\mathbf{R}^2} f(x,z,y) \; dx \; dz. \tag{7.32b}$$

The next aim is to express problem (7.31) in terms of these functions.
To this end, first apply the mean value theorem,

$$\int_{\mathbf{R}^2} b(y)b(z)f(x,y,z) \; dy \; dz = \underbrace{b\left(\eta(x)\right) b\left(\zeta_1(x)\right)}_{=:C_1^{SOI}(x)} \int_{\mathbf{R}^2} f(x,y,z) \; dy \; dz \tag{7.33a}$$

$$\int_{\mathbf{R}^2} a(x)b(z)f(x,y,z) \; dx \; dz = \underbrace{a\left(\xi(y)\right) b\left(\zeta_2(y)\right)}_{=:C_2^{SOI}(y)} \int_{\mathbf{R}^2} f(x,y,z) \; dx \; dz \tag{7.33b}$$

where ξ, η, ζ_1, ζ_2, C_1^{SOI} and C_2^{SOI} are some unknown functions.

Now, the first modelling step comes into play. Approximate C_1^{SOI} and C_2^{SOI} by a model for f. Set

$$f^{mod}(x,y,z) = \begin{cases} \kappa_L^{-11/3} & \text{for } \|(x,y,z)\| < \kappa_L \\ \|(x,y,z)\|^{-11/3} & \text{for } \kappa_L < \|(x,y,z)\| < \kappa_c \\ 0 & \text{for } \|(x,y,z)\| > \kappa_c \end{cases}. \tag{7.34}$$

κ_L denotes the smallest resolved wavenumber of the simulation and κ_c denotes the LES cutoff wavenumber. For $\kappa_L < \|(x,y,z)\| < \kappa_c$, $f^{mod}(x,y,z)$ corresponds to a $\kappa^{-5/3}$ spectrum, $\int_{\|(x,y,z)\|=\kappa} f^{mod}(x,y,z) \sim \kappa^{-5/3}$ (cf. equation (2.19)). It should be noted that the model function f^{mod} assumes isotropy of a single velocity component which does not hold. However, for the present purpose the model is accurate enough. It only serves to estimate C_1^{SOI} and C_2^{SOI}, which are computed from

$$C_1^{SOI}(x) = \frac{\int_{\mathbf{R}^2} b(y)b(z)f^{mod}(x,y,z) \; dy \; dz}{\int_{\mathbf{R}^2} f^{mod}(x,y,z) \; dy \; dz} \quad \text{for } x < \kappa_c \tag{7.35a}$$

$$C_2^{SOI}(y) = \frac{\int_{\mathbf{R}^2} a(x)b(z)f^{mod}(x,y,z) \; dx \; dz}{\int_{\mathbf{R}^2} f^{mod}(x,y,z) \; dx \; dz} \quad \text{for } y < \kappa_c. \tag{7.35b}$$

For $x > \kappa_c$ and $y > \kappa_c$, continue by reflection according to equation (7.12).

Figure 7.9 shows typical samples of transfer functions a and b together with the corresponding functions C_1^{SOI} and C_2^{SOI}. As for ADM, the transfer functions are larger than unity below the cutoff wavenumber κ_c and therefore C_1^{SOI} and C_2^{SOI} increase below κ_c.

Now, implement these estimates in the optimisation problem. Then, the problem is to

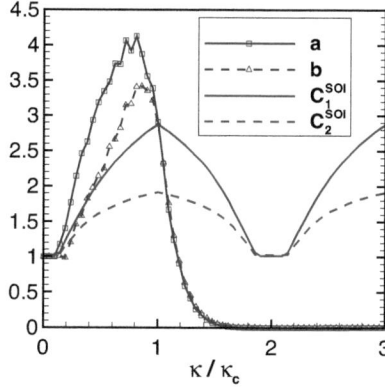

Figure 7.9: Samples of the functions a, b, C_1^{SOI} and C_2^{SOI}. Scaling on wavenumber axis is based on LES cutoff wavenumber κ_c.

find w_l^{na} and w_t^{na} such that $\left\| \begin{pmatrix} E_l^{target}(x) - a(x) C_1^{SOI}(x) E_l^{LES}(x) \\ E_t^{target}(y) - b(y) C_2^{SOI}(y) E_t^{LES}(y) \end{pmatrix} \right\| \to \min .$ (7.36)

It should be noted that C_1^{SOI} depends on b and C_2^{SOI} depends on a. Thus, the problem is two-dimensional.

In order to avoid Fourier transformations, it is desirable to solve problem (7.36) not for w_l^{na} and w_t^{na} but for a and b. This is only possible if $\mathcal{FT}(w_l^{na})$ and $\mathcal{FT}(w_t^{na})$ can be computed from a and b, i.e., if the phase of the Fourier transform is known.

In the present framework the admissibility conditions define the phase. Actually, in section 7.2.4 the stencils will be restricted to be symmetric. This restriction can also be imposed on w_l^{na} and w_t^{na}. Then, $\mathcal{FT}(w_l^{na})$ and $\mathcal{FT}(w_t^{na})$ are real functions, i.e., $\mathcal{FT}(w_l^{na}) = \pm\sqrt{a}$ and $\mathcal{FT}(w_t^{na}) = \pm\sqrt{b}$.

However, the sign is known to be positive because one wants to avoid that an eddy changes its direction by interpolation. Therefore,

$$\mathcal{FT}(w_l^{na}) = +\sqrt{a} \qquad \text{and} \qquad \mathcal{FT}(w_t^{na}) = +\sqrt{b}. \qquad (7.37)$$

Equation (7.37) allows to solve problem (7.36) for a and b. These constraints result from the admissibility conditions, but are not sufficient for admissibility. Therefore, the superscript $\cdot^{na}$ is retained at that point.

As initial conditions, one can set

$$a_0(x) = \frac{E_l^{target}(x)}{E_l^{LES}(x)}, \qquad b_0(x) = \frac{E_t^{target}(x)}{E_t^{LES}(x)}. \qquad (7.38)$$

In order to reduce computational costs, in the present work C_1^{SOI} and C_2^{SOI} were not updated in every iteration of the optimisation. Actually, for most tests, it turned out that it suffices to compute C_1^{SOI} and C_2^{SOI} from a_0 and b_0 and retain these functions for all iterations. This reduces the optimisation problem to two one-dimensional optimisation problems.

In the above steps, the problem was formulated for the continuous functions w_l^{na} and w_t^{na} and their Fourier transforms a and b, respectively. The LES spectra are given in discrete form, and, therefore, the optimisation problem must be solved in discrete form as well. This issue is very simple: The discretised LES spectra E_l^{LES} and E_t^{LES} define a grid on which the target spectrum must be evaluated and on which the terms of problem (7.36) are discretised. This way, one obtains a and b in discretised form.

7.2.4 Imposing admissibility conditions

In the previous section, the optimisation problem was formulated by omitting the admissibility conditions A1 to A4 from page 123. The present section describes a method of constructing an admissible solution (w_l, w_t) from the solution of the unconstrained problem (a, b).

Considering the list on page 123, the solution (a, b) guarantees the properties A5 and A6. The remaining properties are incorporated by designing a compact (property A1) symmetric (property A3, cf. below) interpolation stencil based on cubic splines (property A4). Property A2 is incorporated by normalisation of the stencil.

More precisely, admissibility is realised as follows:

- Compact support (A1):

  w_l and w_t are only admissible if their support is within $]-2, 2[$, i.e., $w_l(x)$ and $w_t(x)$ are zero for $|x| \geq 2$. This leads to a 4-point stencil.

  The inverse Fourier transform of the functions a and b would not possess compact support. This shows that compactness of the support limits the accuracy of the model.

- First-order interpolation (A2):

  The conditions

  $$\sum_{i=-\infty}^{\infty} w_l(x+i) = \sum_{i=-\infty}^{\infty} w_t(x+i) = 1 \tag{7.39}$$

  guarantee first-order interpolation. This means that, given functions $\hat{w}_l$ and $\hat{w}_t$ with

  $$\sum_{i=-\infty}^{\infty} \hat{w}_l(x+i) \neq 0 \quad \text{and} \quad \sum_{i=-\infty}^{\infty} \hat{w}_t(x+i) \neq 0 \quad \text{for any } x \in \mathbf{R}, \tag{7.40}$$

  the functions

  $$w_l(x) = \frac{\hat{w}_l(x)}{\sum_{i=-\infty}^{\infty} \hat{w}_l(x+i)}, \quad w_t(x) = \frac{\hat{w}_t(x)}{\sum_{i=-\infty}^{\infty} \hat{w}_t(x+i)}. \tag{7.41}$$

  define a first-order interpolation stencil.

7 A Novel Particle-LES Model based on Spectrally Optimised Interpolation (SOI)

In the present work, $\hat{w}_l$ and $\hat{w}_t$ are computed from a and b by another optimisation step. Details on this step are given on page 130. Then, the interpolation stencils w_l and w_t are computed from equation (7.41).

- Second-order interpolation in the grid points (A3):

  In general, second-order interpolation means that $u_f(x) = x + \xi$ must be interpolated exactly. ξ can be an arbitrary value. Second-order in the grid points only means that $u_f(x) = x + \xi$ must be interpolated exactly in x_i. In order to obtain this, $\hat{w}_l$ and $\hat{w}_t$ are chosen symmetric around $x = 0$,

  $$\hat{w}_l(x) = \hat{w}_l(-x), \qquad \hat{w}_t(x) = \hat{w}_t(-x). \tag{7.42}$$

  Then, also w_l and w_t are symmetric.

  Second-order accuracy can be shown by interpolating $u_f(x) = \xi + x - x_j$. Respect that constant functions are interpolated exactly and that $x_i = i$. Now, compute $u_{f@p}(x_j)$,

  $$\begin{aligned}
  u_{f@p}(x_j) &= \sum_{i=-\infty}^{\infty} w(x_j - x_i)(\xi + x_i - x_j) = \xi + \sum_{i=-\infty}^{j-1} w(j-i)(i-j) \\
  &+ \sum_{i=j+1}^{\infty} w(j-i)(i-j) = \xi - \sum_{i=1}^{\infty} iw(i) + \sum_{i=1}^{\infty} iw(-i) = \xi \\
  &= u_f(x_j).
  \end{aligned} \tag{7.43}$$

  Therefore the scheme is second-order in the grid points.

  Actually it is even second-order in the intermediate points $(x_j + x_{j+1})/2$ because the interpolation yields for $u_f(x) = x - (x_j + x_{j+1})/2$

  $$\begin{aligned}
  u_{f@p}\left(\frac{x_j + x_{j+1}}{2}\right) &= \sum_{i=-\infty}^{j} w(j + 0.5 - i)(i - j - 0.5) \\
  &+ \sum_{i=j+1}^{\infty} w(j + 0.5 - i)(i - j - 0.5) = 0 = u_f\left(\frac{x_j + x_{j+1}}{2}\right).
  \end{aligned}$$

- Smoothness of the interpolation (A4):

  Taking advantage of symmetry and compact support, the functions $\hat{w}_l$ and $\hat{w}_t$ only need to be specified on the interval $[0, 2[$. In the present work, $\hat{w}_l$ and $\hat{w}_t$ are defined via cubic splines. Each spline is defined on 64 equidistantly placed nodes on the interval $[0, 2[$. Imposing

  $$\hat{w}_{l/t}(-2) = \hat{w}_{l/t}(2) = (\hat{w}_{l/t})'(-2) = (\hat{w}_{l/t})'(2) = 0 \tag{7.44}$$

  as boundary conditions for the spline ensures that $\hat{w}_l$ and $\hat{w}_t$ are continuously differentiable on $\mathbf{R}$ and twice continuously differentiable on $]-2, 2[$. Equation (7.41) shows that this holds for w_l and w_t as well as long as $\sum_{i=-\infty}^{\infty} \hat{w}_l(x+i)$ and $\sum_{i=-\infty}^{\infty} \hat{w}_t(x+i)$

have no roots. Equations (7.8) and (7.27) show that smoothness properties of w_l and w_t transfer to $\mathbf{u}_{f@p}$. This meets property A4.

Remark. Evidently the size of the support of w_l and w_t affects the computational costs of the method. On the other hand, the number of nodes for the splines $\hat{w}_l$ and $\hat{w}_t$ merely affects the overall computational costs (cf. appendix). The choice presented here results in 32 nodes per LES cell and, therefore, makes sense if the LES cell covers 32 Kolmogorov length scales or less. For coarser resolutions, the number of nodes should be increased.

Computation of $\hat{w}_l$ and $\hat{w}_t$

The remaining task is to compute w_l and w_t from a and b. More precisely, this means computing the splines $\hat{w}_l$ and $\hat{w}_t$ on their 64 nodes with boundary conditions (7.44) such that $|\mathcal{FT}(w_l)|^2$ is close to a and $|\mathcal{FT}(w_t)|^2$ is close to b. The exact meaning of 'close' must be defined in this context.

In the present work, 'close' refers to the two-norm. Therefore, a least squares approximation is implemented in two steps. First, a preliminary approximation $\hat{w}_t^p$ and $\hat{w}_l^p$ is computed from a and b by solving the following optimisation problem:

find splines $\hat{w}_t^p$ and $\hat{w}_l^p$ with boundary conditions (7.44) such that

$$\left\| |\mathcal{FT}(w_l^p)(\kappa_i^{res})|^2 - a(\kappa_i^{res}) \right\| \to \min$$

$$\text{and } \left\| |\mathcal{FT}(w_t^p)(\kappa_i^{res})|^2 - b(\kappa_i^{res}) \right\| \to \min \quad (7.45)$$

with

$$w_l^p(x) = \frac{\hat{w}_l^p(x)}{\sum\limits_{i=-\infty}^{\infty} \hat{w}_l^p(x+i)}, \quad w_t^p(x) = \frac{\hat{w}_t^p(x)}{\sum\limits_{i=-\infty}^{\infty} \hat{w}_t^p(x+i)} \quad (7.46)$$

and

$$\kappa_i^{res} \in \{0, \kappa_L, 2\kappa_L, \ldots, 32\kappa_c\}. \quad (7.47)$$

κ^{res} denotes the discretised wavenumbers. As above, κ_L denotes the smallest resolved wavenumber by the LES, and κ_c denotes the LES cutoff wavenumber. The factor 32 stems from the fact that the splines were discretised by 32 nodes per LES cell.

In a second (and final) step, $\hat{w}_t^p$ and $\hat{w}_l^p$ are corrected such that the estimated kinetic energy

7 A Novel Particle-LES Model based on Spectrally Optimised Interpolation (SOI)

is correct. To this end, the following optimisation problem is solved:

find splines $\hat{w}_t$ and $\hat{w}_l$ with boundary conditions (7.44) such that

$$\left| \int_0^\infty E_l^{target}(x)\,\mathrm{d}x - C_0^{SOI}\kappa_L \sum_i |\mathcal{FT}(w_l)(\kappa_i^{res})|^2 \, C_1^{SOI}(\kappa_i^{res}) E_l^{LES}(\kappa_i^{res}) \right| \to \min$$

and

$$\left| \int_0^\infty E_t^{target}(x)\,\mathrm{d}x - C_0^{SOI}\kappa_L \sum_i |\mathcal{FT}(w_t)(\kappa_i^{res})|^2 \, C_2^{SOI}(\kappa_i^{res}) E_t^{LES}(\kappa_i^{res}) \right| \to \min.$$

$\hat{w}_t^p$ and $\hat{w}_l^p$ are used for initialisation. C_1^{SOI} and C_2^{SOI} are computed from a and b, according to equations (7.35a) and (7.35b).

C_0^{SOI} is a model constant. This model constant affects the kinetic energy seen by the particles. It was set on the basis of a single instantaneous LES field such that the kinetic energy seen by inertia-free particles in that field corresponds to the (averaged) DNS kinetic energy.

The kinetic energy of the LES field varies in time. Therefore, the accuracy of the results could be improved by computing C_0^{SOI} from a time-averaged field, but this is not possible in a realistic application where the resolution of the LES is set as high as possible and time averaging is computationally expensive. Therefore, C_0^{SOI} was computed from an instantaneous field.

It should be noted that C_0^{SOI} is the only model constant in SOI. In the present simulations, it ranges from $C_0^{SOI} = 0.98$ to $C_0^{SOI} = 1.13$.

7.2.5 The SOI model formulated as cooking recipe

This section summarises sections 7.2.1 to 7.2.4. The complete model is described again, but no derivations are given.

Pre-processing. Compute the interpolation stencil in a pre-processing step:

1. Define longitudinal and transverse target spectra E_l^{target} and E_t^{target} from the model spectrum proposed by Pope (2000), for example.

2. Compute the longitudinal and transverse LES spectra E_l^{LES} and E_t^{LES} from a single phase LES. Extend these spectra beyond the cutoff wavenumber by reflection, equation (7.12).

3. Solve

$$\left\| \begin{pmatrix} E_l^{target}(x) - a(x)\,C_1^{SOI}(x)\,E_l^{LES}(x) \\ E_t^{target}(y) - b(y)\,C_2^{SOI}(y)\,E_t^{LES}(y) \end{pmatrix} \right\| \to \min \tag{7.48}$$

for the discretised functions a and b. The discretisation grid in the wavenumber space is defined by the LES grid. Compute C_1^{SOI} and C_2^{SOI} from

$$f^{mod}(x,y,z) = \begin{cases} \kappa_L^{-11/3} & \text{for } \|(x,y,z)\| < \kappa_L \\ \|(x,y,z)\|^{-11/3} & \text{for } \kappa_L < \|(x,y,z)\| < \kappa_c \\ 0 & \text{for } \|(x,y,z)\| > \kappa_c \end{cases}, \quad (7.49a)$$

$$C_1^{SOI}(x) = \frac{\int_{\mathbf{R}^2} b(y)b(z)f^{mod}(x,y,z) \, dy \, dz}{\int_{\mathbf{R}^2} f^{mod}(x,y,z) \, dy \, dz} \quad \text{for } x < \kappa_c, \quad (7.49b)$$

$$C_2^{SOI}(y) = \frac{\int_{\mathbf{R}^2} a(x)b(z)f^{mod}(x,y,z) \, dx \, dz}{\int_{\mathbf{R}^2} f^{mod}(x,y,z) \, dx \, dz} \quad \text{for } y < \kappa_c. \quad (7.49c)$$

κ_L denotes the smallest resolved wavenumber of the simulation, and κ_c denotes the LES cutoff wavenumber. For $x > \kappa_c$ and $y > \kappa_c$, continue $C_1^{SOI}(x)$ and $C_2^{SOI}(y)$ by reflection according to equation (7.12).

4. Choose the support size of the interpolation stencil; e.g., for a four point stencil, impose the support (in cell units) $]-2,2[$.

5. Define the stencil using two symmetric cubic splines: one for the longitudinal and one for the transverse direction. Choose the number of nodes for the splines, e.g., 32 nodes per LES cell. Distribute these nodes equidistantly on the positive half support of the stencil, taking advantage of symmetry. Denote the splines by $\hat{w}_l$ and $\hat{w}_t$. Impose the boundary conditions (here formulated for a four point stencil)

$$\hat{w}_{l/t}(-2) = \hat{w}_{l/t}(2) = (\hat{w}_{l/t})'(-2) = (\hat{w}_{l/t})'(2) = 0. \quad (7.50)$$

6. Set the preliminary stencils

$$w_l^p(x) = \frac{\hat{w}_l^p(x)}{\sum\limits_{i=-\infty}^{\infty} \hat{w}_l^p(x+i)}, \quad w_t^p(x) = \frac{\hat{w}_t^p(x)}{\sum\limits_{i=-\infty}^{\infty} \hat{w}_t^p(x+i)} \quad (7.51)$$

and compute the corresponding preliminary splines $\hat{w}_l^p$ and $\hat{w}_t^p$ by solving the optimisation problem (here formulated for a four-point stencil, 32 nodes per LES cell)

find splines $\hat{w}_l^p$ and $\hat{w}_t^p$ with boundary conditions (7.50) such that

$$\left\| |\mathcal{FT}(w_l^p)(\kappa_i^{res})|^2 - a(\kappa_i^{res}) \right\| \to \min$$

$$\text{and } \left\| |\mathcal{FT}(w_t^p)(\kappa_i^{res})|^2 - b(\kappa_i^{res}) \right\| \to \min \quad (7.52)$$

with $\kappa_i^{res} \in \{0, \kappa_L, 2\kappa_L, \ldots, 32\kappa_c\}$. \quad (7.53)

7 A Novel Particle-LES Model based on Spectrally Optimised Interpolation (SOI) 133

7. Use the preliminary splines $\hat{w}_l^p$ and $\hat{w}_t^p$ as initial conditions to solve the following optimisation problem:

find splines $\hat{w}_t$ and $\hat{w}_l$ with boundary conditions (7.50) such that

$$\left| \int_0^\infty E_l^{target}(x)\,\mathrm{d}x - C_0^{SOI} \kappa_L \sum_i |\mathcal{FT}(w_l)(\kappa_i^{res})|^2 \, C_1^{SOI}(\kappa_i^{res}) E_l^{LES}(\kappa_i^{res}) \right| \to \min$$

and

$$\left| \int_0^\infty E_t^{target}(x)\,\mathrm{d}x - C_0^{SOI} \kappa_L \sum_i |\mathcal{FT}(w_t)(\kappa_i^{res})|^2 \, C_2^{SOI}(\kappa_i^{res}) E_t^{LES}(\kappa_i^{res}) \right| \to \min$$

with $w_l(x) = \dfrac{\hat{w}_l(x)}{\sum_{i=-\infty}^{\infty} \hat{w}_l(x+i)}$ and $w_t(x) = \dfrac{\hat{w}_t(x)}{\sum_{i=-\infty}^{\infty} \hat{w}_t(x+i)}$ (7.54)

Set the model constant C_0^{SOI} such that the kinetic energy seen by inertia free particles in a single instantaneous LES field corresponds to the target kinetic energy.

8. Store the cubic splines $\hat{w}_l$ and $\hat{w}_t$ on hard disk.

With these eight pre-processing steps, the interpolation stencils are defined.

At runtime. Compute the fluid velocity seen by the particles from

$$\mathbf{u}_{f@p}(x) = \sum_{i,j,k} \mathbf{W}_{SOI}(\xi_i, \eta_j, \zeta_k) \langle \mathbf{u}_f \rangle (x_i, y_j, z_k) \tag{7.55a}$$

with $\xi_i = x - x_i, \quad \eta_j = y - y_j, \quad \zeta_k = z - z_k$ (7.55b)

and $\mathbf{W}_{SOI}(\xi, \eta, \zeta) = \begin{pmatrix} w_l(\xi) w_t(\eta) w_t(\zeta) & 0 & 0 \\ 0 & w_t(\xi) w_l(\eta) w_t(\zeta) & 0 \\ 0 & 0 & w_t(\xi) w_t(\eta) w_l(\zeta) \end{pmatrix}$

and $w_l(x) = \dfrac{\hat{w}_l(x)}{\sum_{i=-\infty}^{\infty} \hat{w}_l(x+i)}, \quad w_t(x) = \dfrac{\hat{w}_t(x)}{\sum_{i=-\infty}^{\infty} \hat{w}_t(x+i)}.$ (7.55c)

Concerning computational requirements, the CPU time required for the SOI model with a four-point stencil is comparable to the CPU time for fourth-order interpolation. The only difference is that, for SOI, the interpolation weights equal the quotient of two cubic splines (equation (7.55c)), whereas for fourth-order interpolation, the weights equal a cubic polynomial (equation (7.19)). The denominator of equation (7.55c) can be computed very efficiently and, therefore, the number of floating point operations is only 7% higher for SOI than for fourth-order interpolation (cf. appendix).

Furthermore, the appendix contains a comparison of SOI against other particle-LES models in terms of computational costs. The other models are analysed in combination with fourth-order interpolation. According to that analysis, SOI is much less expensive than the stochastic models of chapter 6 and, for sufficiently dilute suspensions, also less expensive than ADM.

7.3 Analytical assessment of the SOI model

In the previous section, the SOI model was introduced. In this section, the accuracy of SOI with respect to the first and second moments listed in chapter 4 is discussed. The present section focuses on a theoretical discussion; the corresponding numerical results are presented in section 7.4.

The present section is the counterpart to section 6.3, which is the analytical assessment of ADM and the stochastic models. In particular, the same assumptions are made; i.e., the particles are assumed to be distributed homogeneously, Stokes drag is assumed to be linear, and the analysis focuses on the structure of the model and not on the parameters. Concerning the last issue, for ADM, it was assumed that the energy spectrum below the cutoff wavenumber is predicted exactly by ADM. Correspondingly, for SOI, it is assumed that the one-dimensional spectra E_l and E_t are reconstructed for all wavenumbers. This is possible if the SOI stencils are chosen large enough.

This means that the energy spectrum function E is reconstructed because (see Pope, 2000)

$$E(\kappa) = -\kappa \frac{\mathrm{d}}{\mathrm{d}\kappa}\left(\frac{1}{2}E_l(\kappa) + E_t(\kappa)\right). \tag{7.56}$$

SOI was developed for isotropic turbulence. In isotropic turbulence, all first moments are zero in the exact solution and in the solution computed with SOI. Therefore, the first moments are predicted correctly by SOI in isotropic turbulence.

Concerning second moments, it was already argued in section 6.3.3 that it is sufficient to analyse $\overline{u_{f@p,i}(\tau)u_{f@p,i}(t)}$ because of equations (4.5b) and (4.7b).

SOI was constructed such that the spectrum seen by the particles is optimised against a model spectrum. This means that if this model spectrum is correct, then the second moment in fluid velocity seen by the particles, $\overline{u_{f@p,i}(t)u_{f@p,i}(t)}$, is predicted correctly by SOI (as noted above, it is assumed that the complete spectrum is reconstructed).

The autocovariance $\overline{u_{f@p,i}(\tau)u_{f@p,i}(t)}$ is a Lagrangian quantity; it equals the Fourier transform of the Lagrangian spectrum (cf. e.g. Pope, 2000). This means that predicting $\overline{u_{f@p,i}(\tau)u_{f@p,i}(t)}$ correctly is equivalent to predicting the Lagrangian spectrum correctly.

For SOI, this is not guaranteed because SOI works on the Eulerian spectrum. If the Eulerian spectra of two space- and time-dependent velocity fields are equal, then this does not necessarily entail equality of the Lagrangian spectra. This can be demonstrated by a simple one-dimensional example. Let the two velocity fields u_1 and u_2 be

$$u_1(x,t) = \sin(x) \qquad u_2(x,t) = \sin(x+t). \tag{7.57}$$

Notably, both fields show the same Eulerian spectra, namely, a Dirac peak at $\kappa = 1$.

Now, initialise inertia free particles equally distributed on the interval $x \in [0, 2\pi]$. Compute the corresponding particle paths for both velocity fields,

$$x_1(t) = x_1(0) + \int_0^t u_1(x_1(\tau),\tau) \, d\tau \qquad x_2(t) = x_2(0) + \int_0^t (x_2(\tau),\tau) \, d\tau. \qquad (7.58)$$

Then, compute the Lagrangian spectra from $u_1(x_1(\tau),\tau)$ and $u_2(x_2(\tau),\tau)$. These are shown in figure 7.10. Notably they are not equal.

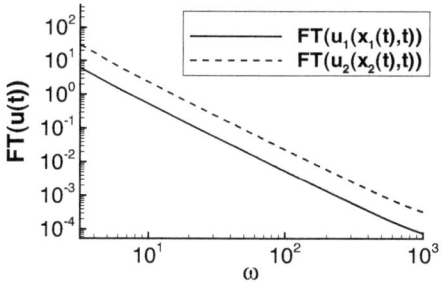

Figure 7.10: Example showing that the equality of Eulerian spectra does not necessarily entail equality of the Lagrangian spectra.

However, turbulence is not an arbitrary flow field. In particular, if Kolmogorov's hypotheses are extended to Lagrangian spectra, then one can derive a model to relate Eulerian spectra to Lagrangian spectra. Corrsin (1963) proposed one of the first of such models, which was later refined by Tennekes & Lumley (1972). Fung et al. (1992) and Lien et al. (1998) extended Corrsin's model by postulating that 'the energy at each wavenumber is spread over a range of frequencies around a characteristic frequency'. The idea behind this model is that a spatial wave generates a temporal Lagrangian wave.

More precisely, Lien et al. (1998) define a distribution $F(\omega, \kappa)$ by

$$F(\omega,\kappa) = \frac{1}{\sqrt{2\pi}\epsilon^{1/3}\kappa^{2/3}} \left(\exp\left(-\frac{(\omega - \epsilon^{1/3}\kappa^{2/3})^2}{2\epsilon^{2/3}\kappa^{4/3}} \right) + \exp\left(-\frac{(\omega + \epsilon^{1/3}\kappa^{2/3})^2}{2\epsilon^{2/3}\kappa^{4/3}} \right) \right). \qquad (7.59)$$

$F(\omega, \kappa)$ quantifies the transfer of energy from wavenumber κ to frequency ω and vice versa. F is shown for fixed κ and for fixed ω in figure 7.11.

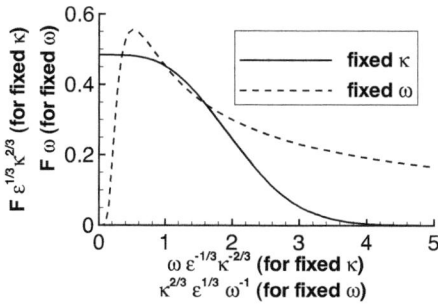

Figure 7.11: The distribution F from equation (7.59) for fixed κ and for fixed ω

With the model of Lien et al. (1998), one can compute the Lagrangian spectrum $E_{Lag}(\omega)$ of an inertia-free particle from the energy spectrum function $E(\kappa)$,

$$E_{Lag}(\omega) = \left\|\mathcal{FT}(\mathbf{u}_{f@p})(\omega)\right\|^2 = \int_0^\infty F(\omega,\kappa)E(\kappa)\,\mathrm{d}\kappa. \tag{7.60}$$

This holds for the DNS spectra. For SOI, the integrand in equation (7.60) can be retained in the range $\kappa < \kappa_c$, i.e., below the cutoff wavenumber, because SOI reconstructs the energy spectrum function. Beyond κ_c, SOI produces a damped reflection of low wavenumbers by construction. For example, for $\kappa_c < \kappa < 2\kappa_c$, SOI reflects the energy at wavenumber $2\kappa_c - \kappa$ to κ. This reflection produces the correct energy spectrum function $E(\kappa)$, but the temporal behaviour of this spatial wave depends on the temporal behaviour of the wave with wavenumber $2\kappa_c - \kappa$. The wave at κ exists exactly as long as the wave at $2\kappa_c - \kappa$. This means that, with SOI, the energy at wavenumbers beyond the cutoff wavenumber is spread over a range of frequencies that is determined by the corresponding resolved wavenumber. For example, for $\kappa_c < \kappa < 2\kappa_c$, the integrand reads $F(\omega, 2\kappa_c - \kappa)E(\kappa)$ instead of $F(\omega,\kappa)E(\kappa)$. Consequently, SOI produces the Lagrangian spectrum

$$E_{Lag}^{SOI}(\omega) = \left\|\mathcal{FT}(\mathbf{u}_{f@p}^{SOI})(\omega)\right\|^2 = \int_0^\infty F(\omega,r(\kappa))E(\kappa)\,\mathrm{d}\kappa \tag{7.61}$$

where r denotes the reflection,

$$r(\kappa) = \min\left\{\kappa \mod 2\kappa_c,\ 2\kappa_c - (\kappa \mod 2\kappa_c)\right\}. \tag{7.62}$$

Here, mod is the modulo operator. Evidently, E_{Lag} and E_{Lag}^{SOI} are not equal, which indicates that the second moments in the particle velocity and particle position are not predicted correctly by SOI. However, the difference between E_{Lag} and E_{Lag}^{SOI} may be small. Figure 7.12 shows E_{Lag} and E_{Lag}^{SOI}, which are both computed from the model spectrum of section 2.1.3 for $Re_\lambda = 265$ and $\kappa_c = \pi/(37.7\eta_K)$. This is the cutoff wavenumber from table 5.2. Figure 7.12 shows that E_{Lag} and E_{Lag}^{SOI} differ only at high frequencies.

Figure 7.12 also shows the Lagrangian spectra for ADM,

$$E_{Lag}^{ADM}(\omega) = \left\|\mathcal{FT}(\mathbf{u}_{f@p}^{SOI})(\omega)\right\|^2 = \int_0^{\kappa_c} F(\omega,\kappa)E(\kappa)\,\mathrm{d}\kappa \tag{7.63}$$

and for LES without particle-LES model,

$$E_{Lag}^{w/o}(\omega) = \left\|\mathcal{FT}(\mathbf{u}_{f@p}^{SOI})(\omega)\right\|^2 = \int_0^{\kappa_c} F(\omega,\kappa)\,(\mathcal{FT}(G))(\kappa)\,E(\kappa)\,\mathrm{d}\kappa. \tag{7.64}$$

The spectrum for ADM (equation (7.63)) is valid under the best case assumptions for ADM from chapter 6. $\mathcal{FT}(G)$ is the filter transfer function of the Lagrangian Smagorinsky model (cf. section 3.1.2). Figure 7.12 shows that the spectrum from SOI is much closer

7 A Novel Particle-LES Model based on Spectrally Optimised Interpolation (SOI)

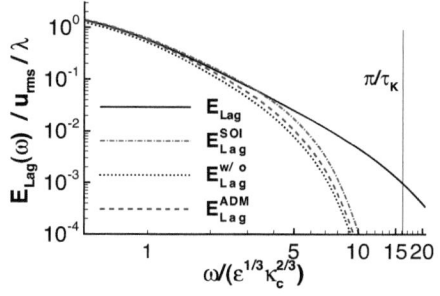

Figure 7.12: Lagrangian spectra computed from the model spectrum of section 2.1.3 for $Re_\lambda = 265$ and equation (7.61). Scaling on the frequency axis is based on the frequency associated with the cutoff wavenumber κ_c. The value for κ_c is $\pi/37.7\eta_K^{-1}$ (cf. table 5.2).

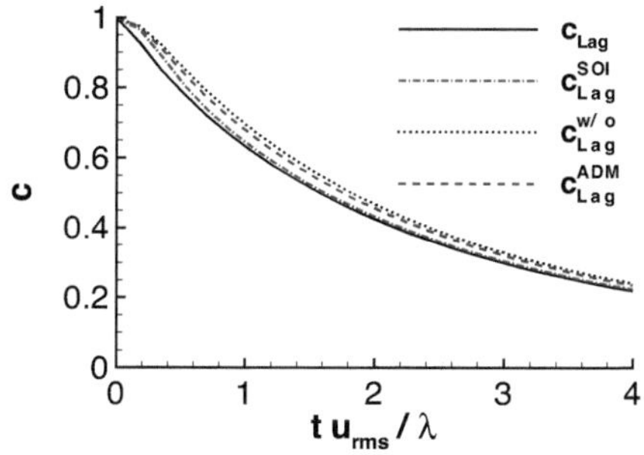

Figure 7.13: Autocorrelation functions that correspond to the spectra plotted in figure 7.12.

to the reference spectrum E_{Lag} than the spectrum from ADM or LES without particle-LES model. This fact demonstrates potentially higher accuracy for SOI than for the other approaches.

The above considerations are only strictly valid for inertia-free particles because Lien et al. (1998) defined the distribution $F(\omega, \kappa)$ for inertia-free particles. For other particles, a similar distribution could be defined that leads to comparable results.

However, the quantity of interest is the autocovariance $\overline{u_{f@p,i}(\tau)u_{f@p,i}(t)}$. It has already been mentioned that, for $\tau = t$, SOI predicts the correct value. For $\tau \neq t$, the autocovariance can be computed from the correlations

$$c(t) = \frac{\overline{u_{f@p,i}(\tau)u_{f@p,i}(\tau+t)}}{\overline{u_{f@p,j}(\tau)u_{f@p,j}(\tau)}} = \mathcal{FT}^{-1}(E_{Lag})(t) \tag{7.65a}$$

$$c^{SOI}(t) = \frac{\overline{u^{SOI}_{f@p,i}(\tau)u^{SOI}_{f@p,i}(\tau+t)}}{\overline{u^{SOI}_{f@p,j}(\tau)u^{SOI}_{f@p,j}(\tau)}} = \mathcal{FT}^{-1}(E^{SOI}_{Lag})(t) \tag{7.65b}$$

$$c^{w/o}(t) = \mathcal{FT}^{-1}(E^{w/o}_{Lag})(t) \tag{7.65c}$$

$$c^{ADM}(t) = \mathcal{FT}^{-1}(E^{ADM}_{Lag})(t). \tag{7.65d}$$

These are plotted in figure 7.13. The figure shows that, for all $t > 0$, the correlations c^{SOI}, $c^{w/o}$ and c^{ADM} are higher than c. This indicates an overprediction of the integral time scale. For $t \lesssim 0.3\lambda/u_{rms}$, the correlations c^{SOI}, $c^{w/o}$ and c^{ADM} are very similar. Beyond this value, c^{SOI} approaches c much faster than the other correlations, which indicates that the overprediction with SOI is the lowest. The integral time scales, i.e., $\int_0^\infty c(t)\,dt$ and $\int_0^\infty c^{SOI}(t)\,dt$, differ only by 2.2%, whereas $\int_0^\infty c(t)\,dt$ and $\int_0^\infty c^{w/o}(t)\,dt$ differ by 7.3% (cf. table 7.1).

Table 7.1: Integral time scales computed from the model presented in section 7.3 for SOI, ADM and LES without particle-LES model.

	reference value	SOI	ADM	w/o particle-LES model
$t_{u@p} u_{rms}/\lambda$	4.50	4.60	4.81	4.83
ratio to reference value	n.a.	1.022	1.069	1.073

To conclude, if the optimisation of the interpolation stencils leads to correct spectra (i.e., if the target spectra are correct and the stencils are sufficiently large), then SOI predicts second moments almost correctly in isotropic turbulence. This means that, in isotropic turbulence, SOI is expected to be advantageous compared to ADM or the stochastic models presented in chapter 6.

7.4 Numerical assessment of the SOI model

The previous section contained an analytical assessment of SOI. In the present section, SOI is assessed using numerical simulation.

The present section corresponds to section 6.4, which describes the numerical assessment of ADM and the stochastic models. In particular, SOI is compared to ADM because ADM was found to be superior to the stochastic models and because SOI is an extension of ADM.

The test case is again forced isotropic turbulence. Unlike in section 6.4, in this section, three Reynolds numbers are considered: $Re_\lambda = 52$, 99 and 265. At all Reynolds numbers, DNS, LES with ADM, LES with SOI and LES without particle-LES model were conducted. At $Re_\lambda = 52$, ADM and SOI were tested with two different target spectra: the DNS spectrum and the model spectrum from section 2.1.3. At $Re_\lambda = 99$ and 265, only the model spectrum was used as target spectrum.

In section 7.4.1, the interpolation stencils of SOI and ADM are presented. Section 7.4.2 shows one-dimensional spectra for ADM and SOI; section 7.4.3 contains statistics on the particle dynamics.

7.4.1 Interpolation stencils

Figure 7.14 shows stencils for ADM and SOI. The ADM stencils were computed as proposed in section 6.4.2, while the SOI stencils were computed as explained above. For each of the four parts of the figure, the input data for ADM and SOI are the same: the LES spectrum and a target spectrum, which is either from DNS or from the model spectrum from section 2.1.3.

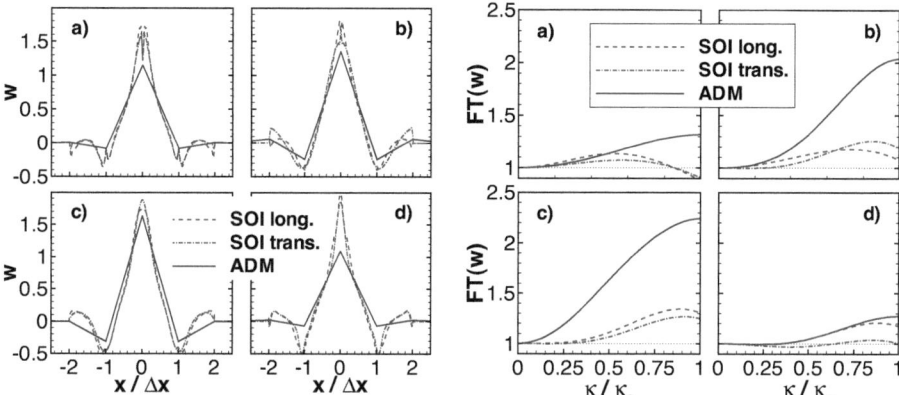

Figure 7.14: Left: Stencils for ADM and SOI. Right: Corresponding transfer functions. Continuous line: ADM stencil, dashed line: longitudinal SOI stencil, dash-dotted line: transverse SOI stencil. a) for $Re_\lambda = 52$ with the DNS spectrum as target spectrum, b) for $Re_\lambda = 52$ with the model spectrum from section 2.1.3 as target spectrum, c) for $Re_\lambda = 99$ with the model spectrum from section 2.1.3 as target spectrum, d) for $Re_\lambda = 265$ with the model spectrum from section 2.1.3 as target spectrum.

In all configurations, the extrema of the SOI stencils are higher than those of the ADM stencils. It is remarkable that all SOI stencils show very steep gradients close to the boundaries $|x/\Delta x| \to 2^-$. The reader is reminded that, by construction, the SOI stencils are all continuously differentiable. In particular, the gradients in $x/\Delta x = \pm 2$ are zero. For the $Re_\lambda = 52$ testcase, one can observe that the steep gradients reoccur at each integer value of $x/\Delta x$ because of normalisation (cf. equations (7.41) and (7.55c)).

Furthermore, all SOI stencils resemble a Mexican hat function $(1 - (x/\sigma)^2)e^{-(x/\sigma)^2/2}$ with $\sigma \approx 0.9\Delta x$. Longitudinal and transverse SOI stencils differ only slightly, which illustrates the isotropy of the LES filter.

Figure 7.14 also shows the transfer functions of these stencils. It is remarkable that all SOI transfer functions attain their maximum below the LES cutoff wavenumber κ_c, which accounts for the slight increase in the transfer function of the fluid-LES model just below κ_c (cf. figure 3.2).

7.4.2 One-dimensional Spectra

The SOI model is based on the reconstruction of one-dimensional spectra. As a result of the admissibility conditions A1 through A4, the model cannot reproduce the one-dimensional spectra exactly. Therefore, it is interesting to analyse the capability of the SOI model to improve one-dimensional spectra.

In figure 7.15, the longitudinal and transverse spectra at $Re_\lambda = 52$ from DNS, LES with ADM and LES with SOI are presented. Here, the DNS spectrum was used for stencil

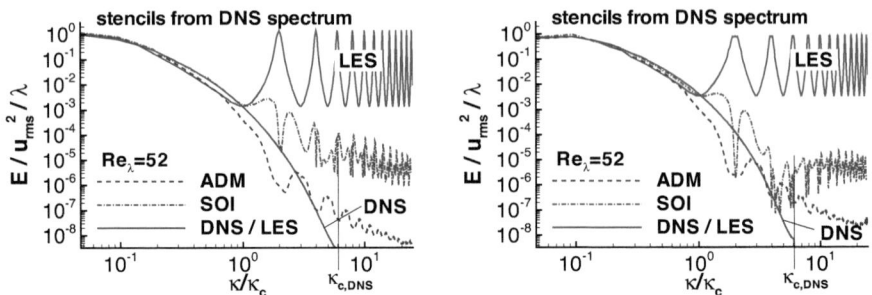

Figure 7.15: Longitudinal (left) and transverse (right) spectra seen by particles in isotropic turbulence at $Re_\lambda = 52$, computed by LES with ADM and SOI. ADM and SOI stencils were obtained by optimisation against the DNS spectrum.[1]

construction for ADM and SOI.

In section 7.1.2, the ADM spectrum was already shown, especially the gap between ADM and DNS at $\kappa < \kappa_c$ that results from interpolation. Figure 7.15 shows that this gap is quasi-non-existent for SOI[1]. On the other hand, SOI produces strong overshoots of the spectra just beyond the LES cutoff wavenumber κ_c. This is because of the low Reynolds number of the test case. At $Re_\lambda = 52$, the inertial subrange is quasi-non-existent. The LES cutoff wavenumber is within the dissipative range. SOI is not capable of producing an interpolation stencil that can enhance the spectrum at low wavenumbers as requested and, at the same time, dampen in the dissipative range. At high Reynolds numbers, the inertial subrange extends to higher wavenumbers, and SOI produces better results, see below.

Now, an analysis of the effect of the target spectrum follows. Figure 7.16 shows the results from $Re_\lambda = 52$ with the model spectrum from section 2.1.3 as the target spectrum. As already observed in chapter 5, the inertial subrange from the model spectrum extends to higher wavenumbers than the DNS spectrum. Chapter 6 showed that, concerning the kinetic energy seen by the particles, ADM based on the model spectrum more closely resembles DNS than ADM based on the DNS spectrum. The findings of the present chapter explain this effect. The overestimation of the DNS result by the model spectrum is compensated by damping due to interpolation.

This does not happen with SOI, and, therefore, the difference between ADM and DNS spectra is smaller than the difference between SOI and DNS spectra. On the other hand, during construction, ADM and SOI were both requested to produce the model spectrum and not the DNS spectrum. The spectrum from SOI is closer to this spectrum than the spectrum from ADM. Thus, the poor performance of SOI in comparison to DNS is not a result of model limitations but rather a result of deviations between the DNS spectrum and

---

[1] Remark for figures 7.15, 7.16, 7.17 and 7.18: The results for ADM were obtained by fourth-order interpolation. For reference, the spectra computed from the grid points in DNS and LES (plus reflections) are also shown (continuous lines). Scaling is with reference to the LES cutoff wavenumber κ_c.

7 A Novel Particle-LES Model based on Spectrally Optimised Interpolation (SOI) 141

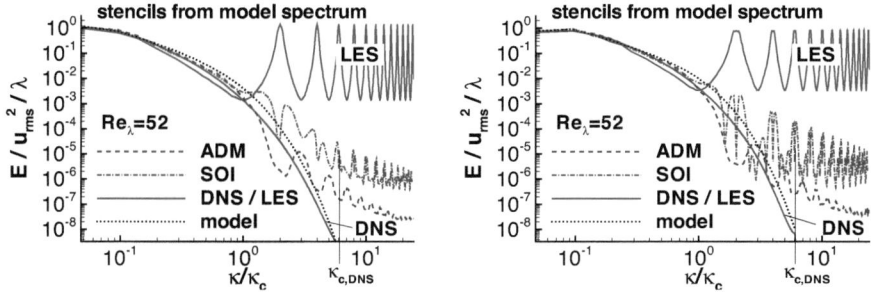

Figure 7.16: Longitudinal (left) and transverse (right) spectra seen by particles in isotropic turbulence at $Re_\lambda = 52$, computed by LES with ADM and SOI. ADM and SOI stencils were obtained by optimisation against the model spectrum (dotted line).[1]

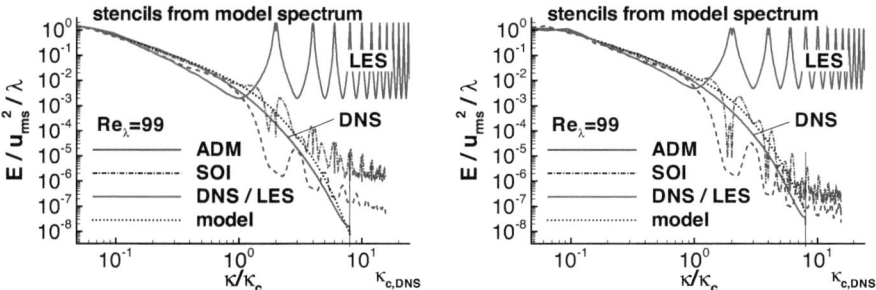

Figure 7.17: Longitudinal (left) and transverse (right) spectra seen by particles in isotropic turbulence at $Re_\lambda = 99$, computed by LES with ADM and SOI. ADM and SOI stencils were obtained by optimisation against the model spectrum (dotted line).[1]

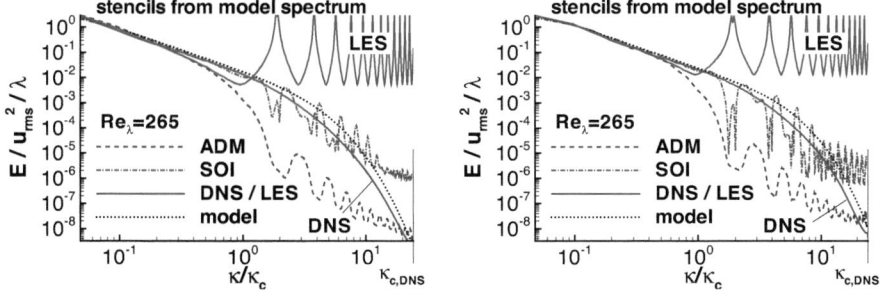

Figure 7.18: Longitudinal (left) and transverse (right) spectra seen by particles in isotropic turbulence at $Re_\lambda = 265$, computed by LES with ADM and SOI. ADM and SOI stencils were obtained by optimisation against the model spectrum (dotted line).[1]

the model spectrum.

The present SOI stencil shows smaller overshoots of the spectra beyond κ_c than the SOI stencil constructed from the DNS spectrum. This is also an effect of the wider inertial subrange of the model spectrum.

Figures 7.17 and 7.18 show results from $Re_\lambda = 99$ and 265. One can observe that, as the Reynolds number increases, performance of SOI improves due to the long inertial subrange. In particular, the overshoots become smaller with higher Reynolds numbers. This is a very promising result that illustrates the accuracy of SOI at very high Reynolds numbers.

7.4.3 Particle dynamics

The previous section concerned the accuracy of SOI with respect to one-dimensional spectra. The present section analyses the accuracy of SOI with respect to particle dynamics, such as kinetic energy, dispersion and preferential concentration. This section corresponds to section 6.4, which is model assessment by numerical simulation.

Again, results from a priori and a posteriori analysis are presented. As in section 6.4, the a priori analysis was conducted in isotropic turbulence at $Re_\lambda = 52$. In the a posteriori analysis, SOI was assessed for the three Reynolds numbers $Re_\lambda = 52, 99$ and 265. All results are summarised at the end of the section in table 7.2.

In the a priori analysis of ADM (section 6.4.2), the particles were traced along paths computed from DNS. The same procedure was followed in the a priori analysis for SOI, which is presented in this section. It is questionable to what extent integral time scales along such a path are relevant. Therefore, in the following, integral time scales are only presented in the a posteriori analysis.

Figure 7.19 shows the kinetic energy seen by the particles. It should be noted that with the proper choice of the model constant C_0^{SOI}, the SOI model could be adapted such that, for one specific Stokes number, the kinetic energy seen by the particles with SOI equals the DNS result. In the present work, C_0^{SOI} was set based on one single instantaneous LES field such that, in that instant, the kinetic energy seen by an inertia-free particle equals the DNS kinetic energy. Because of temporal variations in the resolved turbulent kinetic energy, this method does not lead to a perfect match between SOI and DNS results.

Of course, an ADM stencil can also be constructed such that, for one Stokes number, the kinetic energy seen by the particles matches the DNS result, but this contradicts the idea of the model. With ADM, only wavenumbers up to the cutoff wavenumber are resolved; the small scale energy is unresolved in contrast to SOI. Thus, ADM must always underpredict the kinetic energy seen by the particles.

The results from a priori and a posteriori are in accordance with each other. At $Re_\lambda = 52$, ADM recovers the subgrid kinetic energy well. The ADM stencil that was constructed from the model spectrum leads to higher kinetic energy than the stencil that was constructed from the DNS data, in accordance with the observations from the one-dimensional spectra. Again, in accordance with the one-dimensional spectra, SOI shows still higher kinetic energy and closer resemblance to DNS results than ADM.

The shift along the Stokes number axis is not compensated by SOI. This is not surprising because the generated small scales are generated with high lifetimes (cf. section

7 A Novel Particle-LES Model based on Spectrally Optimised Interpolation (SOI)

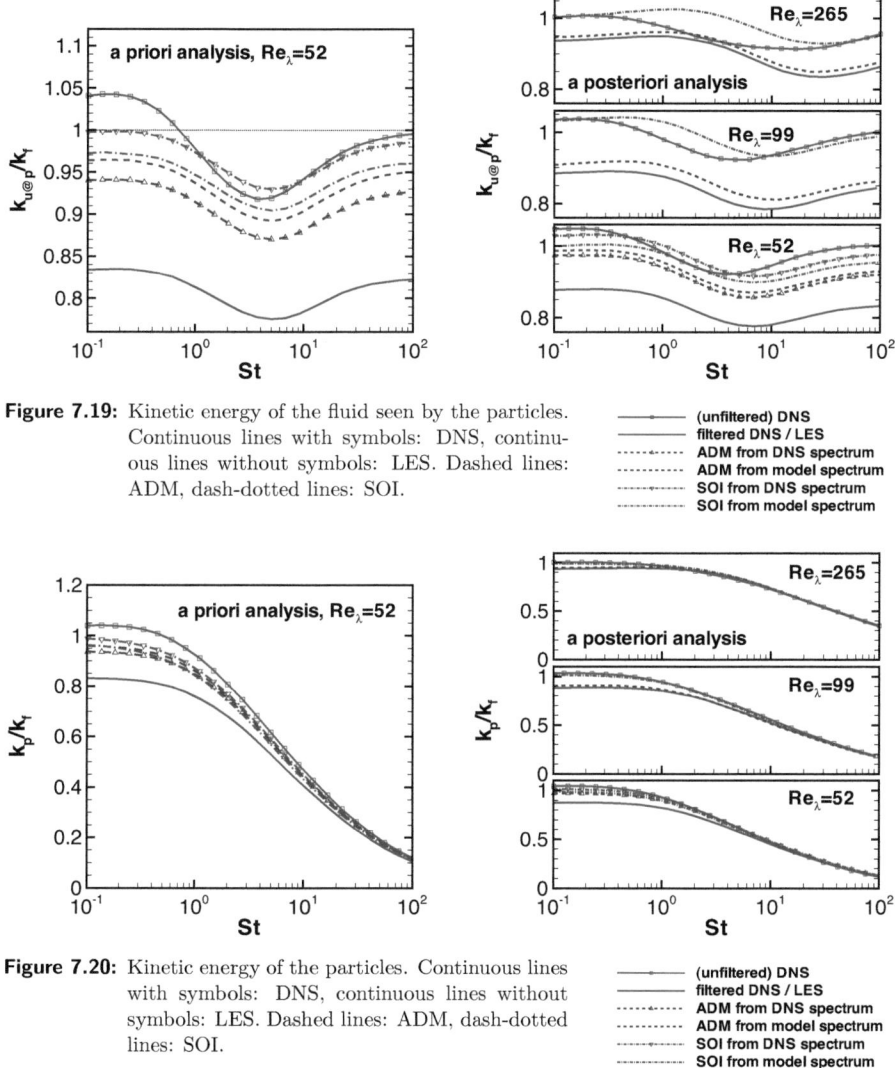

Figure 7.19: Kinetic energy of the fluid seen by the particles. Continuous lines with symbols: DNS, continuous lines without symbols: LES. Dashed lines: ADM, dash-dotted lines: SOI.

— (unfiltered) DNS
— filtered DNS / LES
---- ADM from DNS spectrum
---- ADM from model spectrum
---- SOI from DNS spectrum
---- SOI from model spectrum

Figure 7.20: Kinetic energy of the particles. Continuous lines with symbols: DNS, continuous lines without symbols: LES. Dashed lines: ADM, dash-dotted lines: SOI.

— (unfiltered) DNS
— filtered DNS / LES
---- ADM from DNS spectrum
---- ADM from model spectrum
---- SOI from DNS spectrum
---- SOI from model spectrum

7.3).

At $Re_\lambda = 99$ and 265, the qualitative results are comparable to the $Re_\lambda = 52$ testcase, but the quantitative results differ greatly. At $Re_\lambda = 99$ and 265, ADM shows poor performance. This is expected because the range of unresolved scales is larger at the higher Reynolds

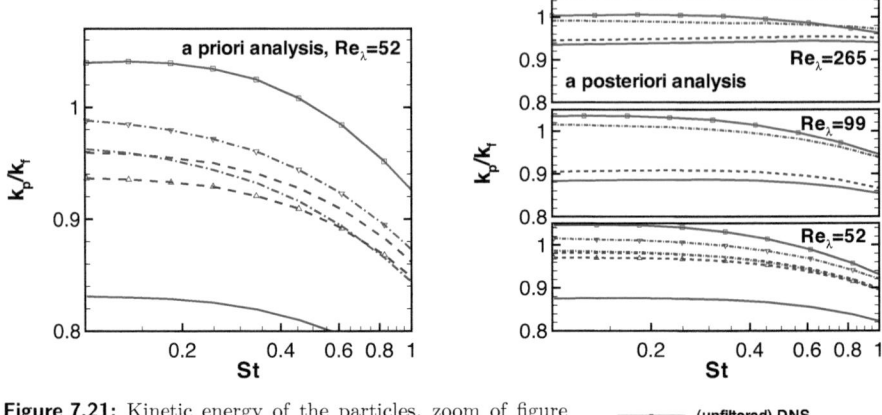

Figure 7.21: Kinetic energy of the particles, zoom of figure 7.20 on $St \leq 10$. Continuous lines with symbols: DNS, continuous lines without symbols: LES. Dashed lines: ADM, dash-dotted lines: SOI.

— (unfiltered) DNS
— filtered DNS / LES
----- ADM from DNS spectrum
----- ADM from model spectrum
-·-·- SOI from DNS spectrum
······ SOI from model spectrum

number. These scales are recovered by SOI but not by ADM, and, therefore, at the higher Reynolds number, SOI shows significantly better results than ADM.

Similar trends can be observed in the kinetic energy of the particles themselves, as illustrated in figures 7.20 and 7.21. Again, the results from a priori and a posteriori analysis are in agreement and again, SOI shows higher accuracy than ADM, in particular at high Reynolds numbers. As expected, the results from DNS, LES, ADM and SOI collapse at high Stokes numbers.

Figures 7.22 and 7.23 show the integral time scales of fluid velocity seen by the particles and particle velocity. At the smallest Reynolds number, SOI performs only slightly better than ADM. This is in accordance with the one-dimensional spectra at $Re_\lambda = 52$ (cf. figures 7.15 and 7.16). It was explained above why the spectral match of SOI improves with a higher Reynolds number.

Accordingly, figures 7.22 and 7.23 show that, at $Re_\lambda = 99$, the match of integral time scales between SOI and DNS is very satisfactory. At this Reynolds number, ADM shows poor performance. In comparison to LES without particle-LES model, ADM merely improves the integral time scale.

At $Re_\lambda = 265$, SOI performs satisfactorily regarding the integral time scale seen by the particles, especially for high Stokes numbers. ADM actually leads to a slightly worse result than LES without particle-LES model. The reason for this is unclear. This may be a statistical artefact.

On the other hand, the integral time scale of the particle velocity at $Re_\lambda = 265$ (figure 7.23), is merely improved by SOI or ADM in comparison to LES without model. However, the results at $Re_\lambda = 265$ are not very reliable because the computational box for these simulations was too small to resolve all scales due to computational limitations (cf. section

7 A Novel Particle-LES Model based on Spectrally Optimised Interpolation (SOI)   145

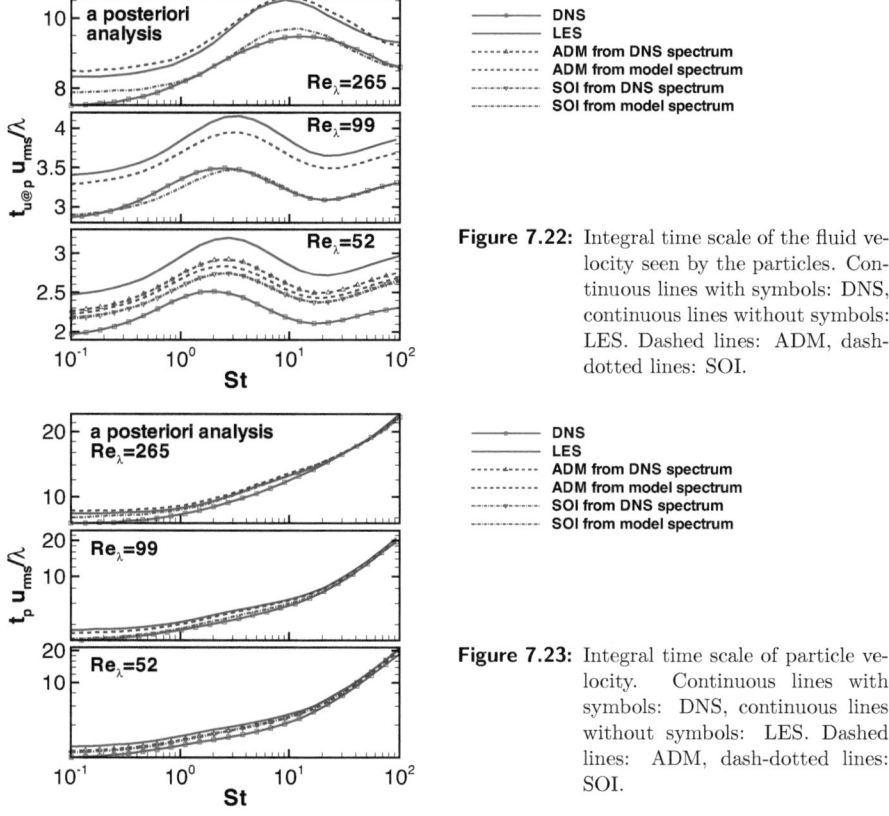

Figure 7.22: Integral time scale of the fluid velocity seen by the particles. Continuous lines with symbols: DNS, continuous lines without symbols: LES. Dashed lines: ADM, dash-dotted lines: SOI.

Figure 7.23: Integral time scale of particle velocity. Continuous lines with symbols: DNS, continuous lines without symbols: LES. Dashed lines: ADM, dash-dotted lines: SOI.

5.2). In particular, a comparison of the predicted time scale from table 7.1 and the DNS result shows that the time scales are far overpredicted in all simulations. However, the analytical assessment showed that integral time scales are overpredicted by SOI. The analytical assessment predicted an overestimation of about 2% for $Re_\lambda = 265$ but figure 7.23 shows that the error is higher than 2%. This is because of a defect in the fluid-LES model which leads to an increase of integral time scales, cf. section 5.5.2. It is not surprising that this defect is very strong at $Re_\lambda = 265$ because at that Reynolds number the filter width is very large.

The rate of dispersion (figure 7.24) is in accordance with the previous results. In particular, SOI performs best at $Re_\lambda = 99$. Again, at $Re_\lambda = 52$, the unsatisfactory spectral match leads to overpredicting the rate of dispersion with SOI, and at $Re_\lambda = 265$ the error from the fluid-LES model becomes apparent.

Concerning preferential concentration, ADM was found to lead to overprediction for $St < 1$ (cf. section 6.4.4). SOI leads to even stronger overprediction in that range (cf. figure 7.25).

7.4 Numerical assessment of the SOI model

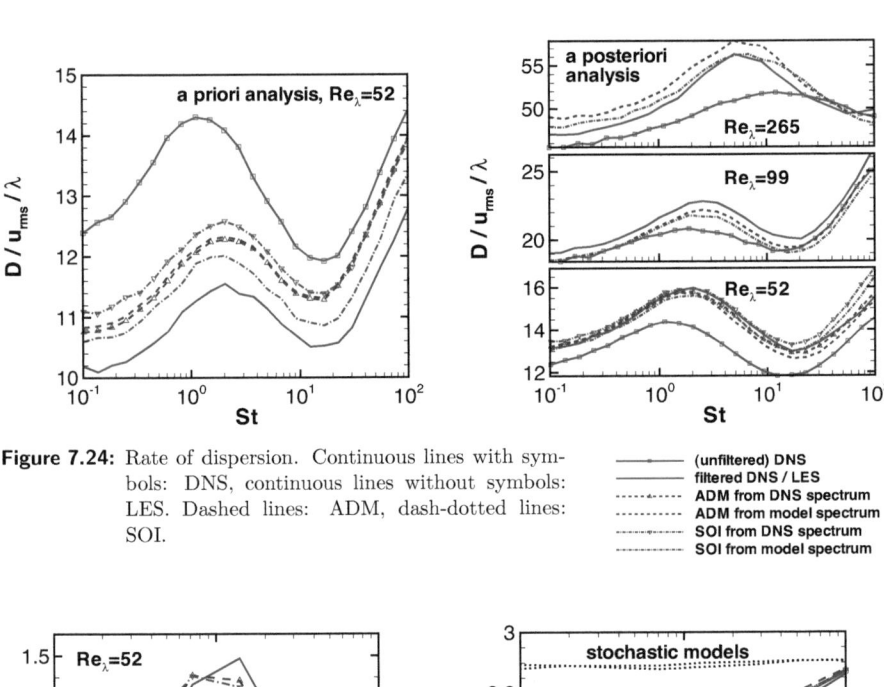

Figure 7.24: Rate of dispersion. Continuous lines with symbols: DNS, continuous lines without symbols: LES. Dashed lines: ADM, dash-dotted lines: SOI.

— (unfiltered) DNS
— filtered DNS / LES
----- ADM from DNS spectrum
······ ADM from model spectrum
—·—· SOI from DNS spectrum
—··— SOI from model spectrum

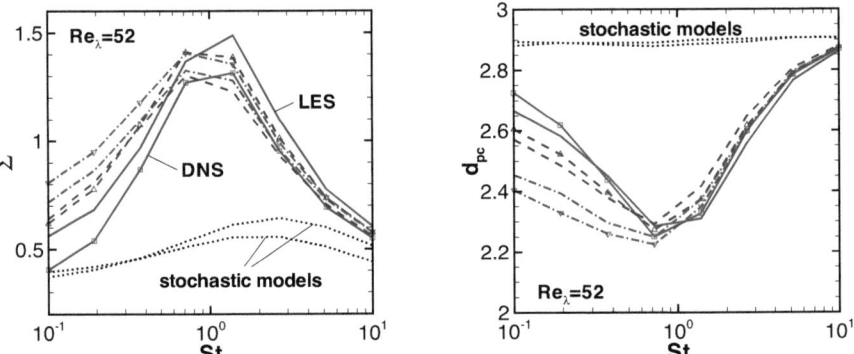

Figure 7.25: Preferential concentration at $Re_\lambda = 52$. Accumulation Σ and fractal dimension d_{pc}. Continuous lines with symbols: DNS, continuous lines without symbols: LES. Dashed lines: ADM, dash-dotted lines: SOI. 'Stochastic models' are the models proposed by Shotorban & Mashayek (2005) and Simonin et al. (1993) (cf. chapter 6)

— DNS
— LES
----- ADM from DNS spectrum
······ ADM from model spectrum
—·—· SOI from DNS spectrum
—··— SOI from model spectrum

This might be because SOI is non-conservative.

In the range $St > 1$, excellent agreement between SOI and DNS can be observed in the accumulation Σ and the fractal dimension d_{pc}. In particular, SOI performs much better with respect to preferential concentration than the stochastic models proposed by Shotorban & Mashayek (2005) and Simonin et al. (1993) (cf. chapter 6).

To conclude, most particle statistics with SOI are closer to DNS than with ADM. In particular, all statistics that are improved by ADM are even more improved by SOI. On the other hand, for the present testcase, preferential concentration is for $St < 1$ better predicted by LES without model than by LES with ADM or SOI. All results are summarised in table 7.2.

Table 7.2: Qualitative summary of the results from section 7.4.3.

	$Re_\lambda = 52$		$Re_\lambda = 99$		$Re_\lambda = 265$	
	ADM	SOI	ADM	SOI	ADM	SOI
$k_{u@p}$	good	very good	poor	excellent	poor	excellent
k_p	good	very good	poor	excellent	poor	excellent
$t_{u@p}$	good	good	poor	excellent	poor	good
t_p	good	good	poor	excellent	poor	poor
D	poor	poor	good	good	poor	poor
Σ, d_{pc}	poor for $St < 1$, excellent for $St > 1$		n.a.	n.a.	n.a.	n.a.

The present results confirm the hypothesis that, as the Reynolds number increases, the performance of ADM worsens. On the other hand, the present results show that SOI is well-suited for high Reynolds number.

In contrast to ADM, the stochastic models of Shotorban & Mashayek (2005) and Simonin et al. (1993) also reconstruct scales that are smaller than the LES grid. From this point of view, the stochastic models are an alternative for SOI at high Reynolds numbers. However, chapter 6 showed that the stochastic models are very unreliable in the sense that the results from stochastic models are often worse than the results without particle-LES model. This holds in particular for preferential concentration. Consequently, SOI is also more accurate than the stochastic models overall. Therefore, SOI can be regarded as the most accurate alternative for particle-LES models at high Reynolds numbers.

7.5 Relation of SOI to other models and outline of extensions for arbitrary flow

In the previous sections, the new model, Spectrally Optimised Interpolation (SOI), was presented and assessed with respect to accuracy and with respect to computational requirements. The present section contains a summary comparison of SOI against other particle-LES models. In particular, it contains a discussion of the relationship between ADM and the two stochastic models that were analysed in chapter 6.

The models are compared against each other with respect to their structure and with respect to their accuracy. Concerning accuracy, the results that were presented in this thesis are summarised. Furthermore, the present section contains two possible extensions of SOI for arbitrary flow because, at present, SOI was only constructed for and applied to forced isotropic turbulence.

SOI can be regarded as an extension of ADM towards kinematic simulation (see e.g. Fung et al., 1992; Malik & Vassilicos, 1999) In kinematic simulation, a velocity field is generated such that its spectrum attains a given model spectrum. Likewise, for SOI, the model spectrum is used as target function for the interpolation stencil.

The basic idea of SOI is similar to the idea of implicit LES proposed by Adams et al. (2004) (see also Hickel et al., 2006, 2008; Hickel & Adams, 2007). Implicit LES stands for numerical discretisation of the Navier–Stokes equations such that the numerical error models small scale effects. Likewise, in SOI, the interpolation stencil is designed such that the interpolation error models the effect of small scale fluctuations on the particles.

In contrast to ADM, SOI also models scales that are smaller than the LES cell width. This is important for most applications that are relevant for LES because, in general, the Reynolds number is so high that the LES cell size is several orders of magnitude larger than the Kolmogorov scale. Such an example was given in the introduction on page 2. Assuming that the flow of the Mississippi river at Baton Rouge must be computed within one week using an 8-core computer that meets current standards, this computation requires an LES cell size of about 13000 times the smallest length scales[2]. At this resolution, the unresolved scales are surely significant.

The stochastic models reconstruct scales smaller than the cell size, but the model assessment in chapter 6 showed that these models have severe deficiencies in the first and second moments and destroy preferential concentration. Thus, these models are often not an option.

SOI, on the other hand, remedies these defects. Small scales are reconstructed, and the results concerning the prediction of second moments and preferential concentration are very promising. Concerning the computational overhead, SOI outperforms the stochastic models and, for sufficiently dilute suspensions, SOI outperforms also ADM, cf. appendix.

One big disadvantage of SOI is that the model was only developed for isotropic turbulence. However, extensions to arbitrary flow are possible, such as using wavelets. Wavelets can be regarded as a localised decomposition of the flow field into its scales. Wavelets were success-

---

[2] Details of the estimate (cf. page 2):
  Available CPU time: $8 \cdot 7 \cdot 24 \cdot 3600 = 5$Mio. CPU seconds
  Length of domain: 18m
  Resolve complete transverse length (inhomogeneous): 1km
  Height of domain: 9m
  Average velocity: $0.9 m/s$
  Cell size: $0.07m \times 0.07m \times 0.07m$, results in $257 \times 14285 \times 129$ cells
  Time step size at $CFL = 1$: $\Delta t = 0.077 s$
  Simulation time: 10 flow through times, 2600 time steps
  Performance: 4 CPU seconds per 10^6 cells per time step
  Overall: 5Mio. CPU seconds
  Smallest scales (see page 2): $5.5 \cdot 10^{-6} m$
  Filter width: $\frac{0.07}{5.5 \cdot 10^{-6}} = 12727$

fully implemented to analyse turbulent flow (see Farge, 1992). These could be a substitute for the Fourier transforms in order to compute the SOI stencil.

Another option would be to substitute the target spectrum with a $\kappa^{-5/3}$ spectrum, i.e., using the spectrum of flow with an infinite Reynolds number and replacing the LES spectrum by the filter transfer function of the LES model multiplied by the target spectrum. Thus, one can obtain an SOI stencil that is independent of the actual LES spectrum. This stencil might serve as a general purpose stencil for arbitrary flow.

7.6 Conclusions of chapter 7

In the present chapter, a novel particle-LES model is proposed. The idea of the model is to take advantage of the numerical error of the interpolation of fluid velocity on particle position such that the spectrum seen by the particles attains a model spectrum.

It was observed that, because of interpolation, the spectrum seen by the particles is not limited by the cutoff wavenumber of the grid. On the contrary, the spectrum may extend to infinite wavenumbers if the interpolation scheme does not damp these wavenumbers sufficiently.

In DNS, this is an undesired property. Beyond the cutoff wavenumber, the spectrum is supposed to be negligible, and, therefore, damping is a desired property in that range.

In LES, the above argument does not hold. The spectrum beyond the cutoff wavenumber is not resolved due to the coarse grid, but the spectral content in this range is not negligible. Therefore, it is questionable whether damping is desired in that range.

It is the task of a particle-LES model to reconstruct this high wavenumber content. Therefore, not-damping can be regarded as modelling.

The second task of a particle-LES model is to enhance the spectrum just below the cutoff wavenumber. Concerning this objective, ADM is the method of choice. Therefore, the new model was constructed as an extension of ADM. With regard to reconstruction of the spectrum beyond the cutoff wavenumber, the new model makes use of the spectral content that falls within that range as a result of the interpolation error. The model was constructed as an interpolation scheme with damping such that the high wavenumber content approaches a model spectrum. Therefore, it is referred to as 'Spectrally Optimised Interpolation' (SOI). The construction presented in this thesis is restricted to isotropic turbulence. Possible extensions for arbitrary flow are only briefly outlined.

It should be mentioned that SOI only reconstructs the high wavenumber spectrum in a statistical sense. SOI was constructed such that the Fourier modes that are generated at wavenumbers beyond the cutoff wavenumber show the correct amplitude on average. The model links the phase of these modes directly to the phase of the corresponding resolved modes. Therefore, all quantities that are related to the phase might not be predicted correctly by SOI.

SOI was assessed with respect to accuracy and with respect to its computational requirements. The assessment with respect to accuracy was performed through analytical considerations and numerical simulation at $Re_\lambda = 52, 99$ and 265. The analytical and numerical results are in agreement with each other. Assessment criteria are statistical moments and preferential concentration (cf. chapter 4). Concerning statistical moments, only second moments were considered because the SOI model was only constructed for isotropic turbulence.

Kinetic energy seen by the particles is predicted correctly by SOI for very small and very high Stokes numbers. For Stokes numbers where clustering affects the kinetic energy seen by the particles, SOI shows the same shift in the Stokes number axis as LES without particle-LES model and LES with ADM.

Concerning kinetic energy of the particles and rate of dispersion, the analytical assessment showed that SOI can be expected to show a slight error, but it also showed that SOI can be expected to perform better than ADM or LES without particle-LES model. The numerical assessment showed that the kinetic energy of the particles is actually predicted significantly better by SOI than by LES with ADM or without particle-LES model.

Preferential concentration was found to be predicted well by SOI for $St > 1$ but overpredicted for $St < 1$. LES with ADM or without particle-LES model also overpredicts preferential concentration, but the effect is strongest with SOI.

In comparison to the stochastic models, SOI was found to be more accurate for all criteria.

According to this study, SOI seems to be the best alternative for applications with high Reynolds numbers. The stochastic models were found to be very inaccurate. At Reynolds numbers that are much higher than the Reynolds numbers considered here, the LES resolution must be very coarse due to computational requirements. In this case, the accuracy of SOI can be expected to increase in comparison to the accuracy of ADM. The presented results confirm this trend.

Concerning computational requirements, SOI was found to require only about 7% more CPU time than LES without particle-LES model and fourth-order interpolation. By comparison, the CPU times for the stochastic models are at least twice as high. For ADM, the computational requirements cannot be compared directly against the requirements for SOI, but for sufficiently dilute suspensions, ADM was found to require more CPU time than SOI.

In conclusion, SOI shows very promising results, and an extension for arbitrary flow might make SOI the particle-LES model of choice.

8 Conclusions

The present thesis concerns Large Eddy Simulation (LES) of particle-laden flow with a focus on the effect of unresolved scales on the particles. This thesis contains new results concerning quantification and modelling of such effects. The corresponding models are referred to as 'particle-LES models'.

8.1 Summary of results

The main findings can be grouped into three parts: analysis of requirements for a particle-LES model (chapter 5), model assessment (chapter 6) and development of a new model (chapter 7).

Part 1: Requirements for a particle-LES model (chapter 5)

The first part (chapter 5) analyses requirements for a particle-LES model, i.e., it contains a quantification of small scale effects. To this end, a series of numerical experiments was conducted. The configurations are forced homogeneous isotropic turbulence at $Re_\lambda = 34$, 52 and 99. Stokes numbers based on the Kolmogorov time scale range from 0.1 to 100. Small scale turbulence was quantified by a priori and a posteriori analysis. Currently, there is no comparable published work that addresses such high Reynolds numbers and such a wide range of Stokes numbers. The motivation for these experiments is to provide insight into physical mechanisms that a particle-LES model must emulate.

With these new data, it was possible to detect that particles tend to cluster in regions where the kinetic energy of the carrier flow is higher or lower than average for the first time. Particles with Stokes numbers greater than one tend to cluster in regions with sub-average kinetic energy, whereas particles with Stokes numbers smaller than one tend to cluster in regions with super-average kinetic energy. Furthermore, the data provided enough information to formulate scaling laws for the Stokes number dependence of kinetic energy seen by the particles and the corresponding integral time scale. The analyses further showed that neglecting subgrid turbulence effects on particle transport merely affects the rate of particle dispersion, as long as the LES is very well resolved. This result is in accordance with previously published results. On the other hand, the new data with high Reynolds numbers showed that, in coarse LES, neglecting subgrid scale turbulence leads to overprediction of the rate of dispersion. This finding is based on the LES model proposed by Meneveau *et al.* (1996). For other models, a similar behaviour is expected.

Altogether, the first part of this thesis illustrates that the effect of the subgrid scale turbulence on the particles cannot be neglected. A corresponding model (particle-LES model) must increase kinetic energy seen by the particles and decrease the corresponding integral time scale. Furthermore, a particle-LES model must preserve preferential concentration.

Part 2: Model assessment (chapter 6)

The second part of this thesis (chapter 6) analyses three of the most commonly used particle-LES models: the Approximate Deconvolution Method (ADM) and two stochastic Langevin-based models. The application of ADM to particle-laden flow can be attributed to Kuerten (2006b). The stochastic models were proposed by Shotorban & Mashayek (2006) and Simonin et al. (1993).

The three models were analysed by analytical and numerical means. The analytical computations are relevant for arbitrary turbulent flow and the numerical simulations are relevant for isotropic turbulence. Where applicable, the analytical results are in good agreement with the numerical results. Both assessment methods provide substantial new insight in sources for model errors.

In particular, the stochastic models show an error in the first and second statistical moments of particle position and velocity for general turbulent flow. The error was found to result from a structural deficiency of the models that had not been discovered yet. Furthermore, the analytical computations demonstrate that the stochastic models can predict kinetic energy and rate of dispersion correctly if the Stokes number is sufficiently small and if the model parameters are chosen optimally. In accordance with that finding, the numerical results illustrate that, for small Stokes numbers, kinetic energy and rate of dispersion are predicted acceptably well, but for high Stokes numbers, the stochastic models lead to unacceptably high errors. With high Stokes numbers, not-modelling showed better results than stochastic modelling.

Concerning ADM, analytical and numerical results demonstrate that the model improves first and second statistical moments of particle position and velocity. However, the results also show that for realistic applications where only coarse LES is possible, the improvement is very small. The reason for this is that ADM cannot generate scales smaller than the LES grid. This is a known conceptual restriction of ADM. However, the analytical computations of the present thesis prove for the first time that this restriction leads to a significant error in the rate of dispersion at all Stokes numbers. The numerical results support this statement.

Furthermore, the numerical results illustrate that preferential concentration is preserved by ADM but destroyed by the stochastic models. To conclude, the stochastic models were found to deliver unsatisfactory results. ADM, on the other hand, was found to deliver acceptable results at the analysed Reynolds number. However, for simulations with a high Reynolds number, only coarse LES is possible. In this case, ADM leads to little improvement because of the conceptual restriction mentioned above. Therefore, a new particle-LES model is necessary that overcomes this restriction.

Part 3: Development of a new model (chapter 7)

In the third part (chapter 7), a new particle-LES model is proposed. The idea behind the model is to identify small scale effects with numerical errors that occur at interpolation of fluid velocity at the particle position in a statistical sense. The model consists of an

8 Conclusions

interpolation scheme that is constructed such that the spectrum seen by the particles approximates a model spectrum. The focus of this interpolation scheme is not on the order of interpolation but on smoothness and the spectral properties of the fluid velocity seen by the particles.

In the present thesis, the model is constructed for application in isotropic turbulence. It is assessed by the same analytical and numerical means that were applied to assess the models mentioned above. The numerical simulations were conducted at $Re_\lambda = 52$, 99 and 265. For reference, ADM was also assessed at these Reynolds numbers. The new model shows very promising results. In particular, the model shows high accuracy in coarse LES with a high Reynolds number, in contrast to ADM.

8.2 Possible extensions of this thesis

The proposed model was developed for isotropic turbulence only. In this thesis, possible extensions for general configurations are briefly outlined but not worked out in detail. With these extensions, the new model might become the particle-LES model of choice. Therefore, it is very desirable to continue model development in that direction.

For model refinement, one could allow for an unsteady interpolation stencil. Then, one could decouple the phase of the resolved and modelled scales. The analytical computations from section 7.3 demonstrate that such a measure could improve time scale prediction. It would be desirable to design the decoupling such that the scaling laws from chapter 5 are recovered.

Other possible extensions of the present thesis concern further analysis of physical mechanisms in particle-laden flow. In chapter 5, a universal subrange of the kinetic energy seen by the particles is postulated. It remains to prove its existence by DNS or experiments with higher Reynolds numbers. With current computers, such a DNS is not possible, but experiments could fill this gap.

Furthermore, there are few reference data available from experiments of particle-laden isotropic turbulence at the absence of gravity. The absence of gravity is important in order to differentiate between the effect of gravity and the effect of turbulence. With such experiments the provided numerical data could be validated and extended towards higher Reynolds number.

Finally, there are still modelling issues for DNS of particle-laden flow if the particles are larger than the Kolmogorov length or as soon as two- or four-way coupling comes into play. These issues are very challenging because they involve a multitude of skills. High performance computing is necessary in order to obtain reference data, and new ideas for modelling must be formulated that require a deep understanding of the involved physical mechanisms. In order to solve these issues, experience with computing, data generation and data analysis; deep knowledge of physics in particle-laden flow, and additionally mathematical and engineering skills for the formulation of an efficient model are necessary. In my opinion, such issues can only be tackled by an interdisciplinary team. Therefore, I want to close this thesis with an appeal for closer collaboration among different disciplines.

Bibliography

ACHENBACH, E. 1974 Vortex shedding from spheres. *J. Fluid Mech.* **62** (02), 209–221.

ADAMS, N. A., HICKEL, S. & FRANZ, S. 2004 Implicit subgrid-scale modeling by adaptive deconvolution. *J. Comput. Phys.* **200** (2), 412–431.

ALMEIDA, T. & JABERI, F. 2008 Large-eddy simulation of a dispersed particle-laden turbulent round jet. *Int. J. Heat Mass Transfer* **51** (3-4), 683–695.

AMIRI, A., HANNANI, S. & MASHAYEK, F. 2006 Large-eddy simulation of heavy-particle transport in turbulent channel flow. *Numer. Heat Transfer, Part B* **50** (4), 285–313.

APTE, S., MAHESH, K. & LUNDGREN, T. 2008 Accounting for finite-size effects in simulations of disperse particle-laden flows. *Int. J. Multiphase Flow* **34** (3), 260–271.

ARMENIO, V. & FIOROTTO, V. 2001 The importance of the forces acting on particles in turbulent flows. *Phys. Fluids* **13** (8), 2437–2440.

ARMENIO, V., PIOMELLI, U. & FIOROTTO, V. 1999 Effect of the subgrid scales on particle motion. *Phys. Fluids* **11** (10), 3030–3042.

AYYALASOMAYAJULA, S., GYLFASON, A., COLLINS, L. R., BODENSCHATZ, E. & WARHAFT, Z. 2006 Lagrangian measurements of inertial particle accelerations in grid generated wind tunnel turbulence. *Phys. Rev. Lett.* **97** (14), 144507.

BALACHANDAR, S. & EATON, J. K. 2010 Turbulence dispersed multiphase flow. *Ann. Rev. Fluid Mech.* **42** (1).

BALACHANDAR, S. & MAXEY, M. R. 1989 Methods for evaluating fluid velocities in spectral simulations of turbulence. *J. Comput. Phys.* **83** (1), 96–125.

BARTEL, A. & GÜNTHER, M. 2002 A multirate W-method for electrical networks in state-space formulation. *J. Comput. Appl. Math.* **147** (2), 411–425.

BASSET, A. B. 1888 *Treatise on Hydrodynamics*. London: Deighton Bell.

BATCHELOR, G. K. 1982 *The Theory of Homogeneous Turbulence*. Cambridge: Cambridge University Press.

BEC, J., BIFERALE, L., CENCINI, M., LANOTTE, A., MUSACCHIO, S. & TOSCHI, F. 2007 Heavy particle concentration in turbulence at dissipative and inertial scales. *Phys. Rev. Lett.* **98** (8).

BEISHUIZEN, N., NAUD, B. & ROEKAERTS, D. 2007 Evaluation of a modified Reynolds stress model for turbulent dispersed two-phase flows including two-way coupling. *Flow. Turbul. Combust.* **79** (3), 321–341.

BERLEMONT, A., DESJONQUERES, P. & GOUESBET, G. 1990 Particle Lagrangian simulation in turbulent flows. *Int. J. Multiphase Flow* **16** (1), 19–34.

BERROUK, A. S., LAURENCE, D., RILEY, J. J. & STOCK, D. E. 2007 Stochastic modelling of inertial particle dispersion by subgrid motion for LES of high Reynolds number pipe flow. *J. Turbul.* **8**, N50.

BIFERALE, L., BOFFETTA, G., CELANI, A., DEVENISH, B. J., LANOTTE, A. & TOSCHI, F. 2004 Multifractal statistics of Lagrangian velocity and acceleration in turbulence. *Phys. Rev. Lett.* **93** (6), 064502.

BINI, M. & JONES, W. P. 2007 Particle acceleration in turbulent flows: A class of nonlinear stochastic models for intermittency. *Phys. Fluids* **19** (3), 035104.

BINI, M. & JONES, W. P. 2008 Large-eddy simulation of particle-laden turbulent flows. *J. Fluid Mech.* **614**, 207–252.

BOGUCKI, D., DOMARADZKI, J. A. & YEUNG, P. K. 1997 Direct numerical simulations of passive scalars with Pr>1 advected by turbulent flow. *J. Fluid Mech.* **343**, 111–130.

BOUSSINESQ, J. 1877 Théorie de l'Écoulement tourbillant. *Acad. Sci. Inst. Fr.* **23**, 46–50.

BOUSSINESQ, J. 1903 *Théorie analytique de la chaleur*. Paris: Gauthier-Villars.

BRENNEN, C. E. 2005 *Fundamentals of Multiphase Flow*, illustrated edition edn. Cambridge: Cambridge University Press.

BRUN, C., JUVÉ, D., MANHART, M. & MUNZ, C. D., ed. 2009 *Numerical Simulation of Turbulent Flows and Noise Generation: Results of the DFG/CNRS Research Groups FOR 507 and FOR 508*, 1st edn. New York: Springer.

BUEVICH, Y. A. 1966 Motion resistance of a particle suspended in a turbulent medium. *Fluid Dyn.* **1** (6), 119.

CHEN, L., GOTO, S. & VASSILICOS, J. C. 2006 Turbulent clustering of stagnation points and inertial particles. *J. Fluid Mech.* **553**, 143–154.

CLIFT, R., GRACE, J. R. & WEBER, M. E. 1978 *Bubbles, Drops and Particles*. New York: Academic Press.

COLLINS, L. R. & KESWANI, A. 2004 Reynolds number scaling of particle clustering in turbulent aerosols. *New J. Phys.* **6**, 119.

CORRSIN, S. 1963 Estimates of the relations between Eulerian and Lagrangian scales in large Reynolds number turbulence. *J. Atmos. Sci.* **20** (2), 115–119.

CORRSIN, S. & LUMLEY, J. 1956 On the equation of motion for a particle in turbulent fluid. *Appl. Sci. Res.* **6**.

CROWE, C., SOMMERFELD, M. & TSUJI, Y. 1998 *Multiphase flows with droplets and particles*. Boca Raton: CRC Press.

DEUFLHARD, P. & BORNEMANN, F. 2008 *Numerische Mathematik 2: Gewöhnliche Differentialgleichungen*, 3rd edn. Berlin: Walter de Gruyter.

DEUFLHARD, P. & HOHMANN, A. 2008 *Numerische Mathematik 1*, 4th edn. Walter de Gruyter.

EGOLF, D. A. & GREENSIDE, H. S. 1994 Relation between fractal dimension and spatial correlation length for extensive chaos. *Nature* **369** (6476), 129–131.

ELGHOBASHI, S. 1991 Particle-laden turbulent flows: Direct simulation and closure models. *Appl. Sci. Res.* **48** (3), 301–314.

ESWARAN, V. & POPE, S. B. 1988 An examination of forcing in direct numerical simulations of turbulence. *Comput. Fluids* **16** (3), 257–278.

EYINK, G. L. & SREENIVASAN, K. R. 2006 Onsager and the theory of hydrodynamic turbulence. *Rev. Mod. Phys.* **78** (1), 87–135.

FARGE, M. 1992 Wavelet transforms and their applications to turbulence. *Ann. Rev. Fluid Mech.* **24** (1), 395–458.

FAXEN, H., WIMAN, A. & OSEEN, C. W. 1922 Die Bewegung einer starren Kugel längs der Achse eines mit zäher Flüssigkeit gefüllten Rohres. *Arkiv för Mathematik, Astronomi och Fysik.* **17** (27).

FEDE, P. & SIMONIN, O. 2006 Numerical study of the subgrid fluid turbulence effects on the statistics of heavy colliding particles. *Phys. Fluids* **18** (4), 045103.

FEDE, P., SIMONIN, O., VILLEDIEU, P. & SQUIRES, K. D. 2006 Stochastic modeling of the subgrid fluid velocity fluctuation seen by inertial particles. In *Proceedings of the Summer Program*. Center for Turbulence Research, Stanford University.

FERZIGER, J. H. & PERIC, M. 1999 *Computational Methods for Fluid Dynamics*, 2nd edn. New York: Springer.

FESSLER, J. R., KULICK, J. D. & EATON, J. K. 1994 Preferential concentration of heavy particles in a turbulent channel flow. *Phys. Fluids* **6** (11), 3742–3749.

FÉVRIER, P., SIMONIN, O. & SQUIRES, K. D. 2005 Partitioning of particle velocities in gas-solid turbulent flows into a continuous field and a spatially uncorrelated random distribution: Theoretical formalism and numerical study. *J. Fluid Mech.* **533**, 1–46.

FOX, R. O. & YEUNG, P. K. 2003 Improved Lagrangian mixing models for passive scalars in isotropic turbulence. *Phys. Fluids* **15** (4), 961–985.

FREUND, R. W. & HOPPE, R. H. W. 2007 *Stoer/Bulirsch: Numerische Mathematik 1*, 10th edn. New York: Springer.

FRISCH, U. 1995 *Turbulence: The Legacy of A. N. Kolmogorov*. Cambridge: Cambridge University Press.

FUKAGATA, K. 1998 Force balance in a turbulent particulate channel flow. *Int. J. Multiphase Flow* **24** (6), 867–887.

FUNG, J. C. H., HUNT, J. C. R., MALIK, N. A. & PERKINS, R. J. 1992 Kinematic simulation of homogeneous turbulence by unsteady random fourier modes. *J. Fluid Mech.* **236**, 281–318.

GEORGE, W. K. 1989 The self preservation of turbulent flows and its relation to initial conditions and coherent structure. In *Advances in Turbulence* (ed. W. K. George & R. Arndt). New York: Hemisphere.

GERMANO, M. 1992 Turbulence: the filtering approach. *J. Fluid Mech.* **238**, 325–336.

GERMANO, M., PIOMELLI, U., MOIN, P. & CABOT, W. H. 1991 A dynamic subgrid-scale eddy viscosity model. *Phys. Fluids A-Fluid* **3** (7), 1760–1765.

GEURTS, B. & VREMAN, B. 2006 Dynamic self-organization in particle-laden channel flow. *Int. J. Heat Fluid Flow* **27** (5), 945–954.

GEURTS, B. J., CLERCX, H. & UIJTTEWAAL, W. 2007 *Particle-Laden Flow: From Geophysical to Kolmogorov Scales*, 1st edn. New York: Springer.

GICQUEL, L. Y. M., GIVI, P., JABERI, F. A. & POPE, S. B. 2002 Velocity filtered density function for large eddy simulation of turbulent flows. *Phys. Fluids* **14** (3), 1196–1213.

GITTERMAN, M. & STEINBERG, V. 1980 Memory effects in the motion of a suspended particle in a turbulent fluid. *Phys. Fluids* **23** (11), 2154–2160.

GOBERT, C., LINK, O., MANHART, M. & ZANKE, U. 2010 Discussion of "Coherent structures in the flow field around a circular cylinder with scour hole" by G. Kirkil, S. G. Constaninescu and R. Ettema. *to appear in J. Hydraul. Eng.* **136** (1).

GOBERT, C. & MANHART, M. 2007 Modeling particle dispersion in LES computations: The effect of the subgrid scale turbulence. In *Proceedings of the 6th International Conference on Multiphase Flow*. Leipzig.

GOBERT, C. & MANHART, M. 2009 Numerical experiments for quantification of small scale effects in particle laden turbulent flow. In *High Performance Computing in Science and Engineering, Garching 2009*. New York: Springer.

GOBERT, C., MOTZET, K. & MANHART, M. 2007 A stochastic model for large eddy simulation of a particle-laden turbulent flow. In *Particle-Laden Flow: From Geophysical to Kolmogorov Scales*, 1st edn. New York: Springer.

GOBERT, C., SCHWERTFIRM, F. & MANHART, M. 2006 Lagrangian scalar tracking for laminar micromixing at high Schmidt numbers. In *Proceedings of the 2006 ASME Joint U.S.-European Fluids Engineering Summer Meeting*. Miami, Fl.

GOTO, S. & VASSILICOS, J. C. 2006 Self-similar clustering of inertial particles and zero-acceleration points in fully developed two-dimensional turbulence. *Phys. Fluids* **18** (11), 115103.

GOTTWALD, B. & WANNER, G. 1981 A reliable Rosenbrock integrator for stiff differential equations. *Computing* **26** (4), 355–360.

GRASSBERGER, P. & PROCACCIA, I. 1983 Measuring the strangeness of strange attractors. *Physica D* **9** (1-2), 189–208.

GUHA, A. 2008 Transport and deposition of particles in turbulent and laminar flow. *Ann. Rev. Fluid Mech.* **40** (1), 311–341.

GUI, N., FAN, J. & CEN, K. 2008 Effect of particle-particle collision in decaying homogeneous and isotropic turbulence. *Phys. Rev. E: Stat. Nonlinear Soft Matter Phys.* **78** (4), 046307.

HAIRER, E. & WANNER, G. 1990 *Solving Ordinary Differential Equations II. Stiff and Differential-Algebraic Problems*. New York: Springer.

HAWORTH, D. C. & POPE 1986 A generalized Langevin model for turbulent flows. *Phys. Fluids* **29** (2), 387–405.

HEINZ, S. 2003 On Fokker-Planck equations for turbulent reacting flows. Part 2. Filter density function for large eddy simulation. *Flow Turbul. Combust.* **70**, 153–181.

HEISENBERG, W. 1948 Zur statistischen Theorie der Turbulenz. *Z. Phys. A-Hadron. Nucl.* **124** (7), 628–657.

HICKEL, S. & ADAMS, N. 2007 On implicit subgrid-scale modeling in wall-bounded flows. *Phys. Fluids* **19** (10), 105106.

HICKEL, S., ADAMS, N. & DOMARADZKI, J. 2006 An adaptive local deconvolution method for implicit LES. *J. Comput. Phys.* **213** (1), 413–436.

HICKEL, S., KEMPE, T. & ADAMS, N. 2008 Implicit large-eddy simulation applied to turbulent channel flow with periodic constrictions. *Theor. Comp. Fluid Dyn.* **22** (3), 227–242.

HINZE, J. O. 1975 *Turbulence*, 2nd edn. New York: McGraw-Hill.

HOGAN, R. C. & CUZZI, J. N. 2001 Stokes and Reynolds number dependence of preferential particle concentration in simulated three-dimensional turbulence. *Phys. Fluids* **13**, 2938–2945.

HWANG, W. & EATON, J. K. 2006 Turbulence attenuation by small particles in the absence of gravity. *Int. J. Multiphase Flow* **32** (12), 1386–1396.

IJZERMANS, R. H. A., HAGMEIJER, R. & VAN LANGEN, P. J. 2007 Accumulation of heavy particles around a helical vortex filament. *Phys. Fluids* **19** (10), 107102.

ISSA, R. & OLIVEIRA, P. 1997 Assessment of a particle-turbulence interaction model in conjunction with an eulerian two-phase flow formulation. In *2nd International Symposium on Turbulence, Heat and Mass Transfer* (ed. K. Hanjalic & T. Peeters), pp. 759–770. Delft: Delft University Press.

JOHANSSON, P. B. V., GEORGE, W. K. & GOURLAY, M. J. 2003 Equilibrium similarity, effects of initial conditions and local Reynolds number on the axisymmetric wake. *Phys. Fluids* **15** (3), 603–617.

KAPS, P. & RENTROP, P. 1979 Generalized Runge-Kutta methods of order four with stepsize control for stiff ordinary differential equations. *Numer. Math.* **33** (1), 55–68.

KINCAID, D. R. & CHENEY, E. W. 2001 *Numerical Analysis: Mathematics of Scientific Computing*, 3rd edn. Pacific Grove, CA: Brooks/Cole.

KLOEDEN, P. E. & PLATEN, E. 2000 *Numerical Solution of Stochastic Differential Equations*, corrected edn. New York: Springer.

KOLMOGOROV, A. N. 1941 The local structure of turbulence in incompressible viscous fluid for very large Reynolds numbers. *Dokl. Akad. Nauk SSSR* **30** (4).

KOLMOGOROV, A. N. 1991 The local structure of turbulence in incompressible viscous fluid for very large Reynolds numbers (translated by V. Levin). *Proc. R. Soc. London, Ser. A* **434** (1890), 9–13.

KRAICHNAN, R. H. 1959 The structure of isotropic turbulence at very high Reynolds numbers. *J. Fluid Mech.* **5** (04), 497–543.

KUBIK, A. & KLEISER, L. 2004 Forces acting on particles in separated wall-bounded shear flow. *Proc. Appl. Math. Mech.* **4**, 512–513.

KUERTEN, J. G. M. 2006a Large-eddy simulation of particle-laden channel flow. In *Direct and Large-Eddy Simulation VI* (ed. E. Lamballais, R. Friedrich, B. J. Geurts & O. Metais). New York: Springer.

KUERTEN, J. G. M. 2006b Subgrid modeling in particle-laden channel flow. *Phys. Fluids* **18**, 025108.

KUERTEN, J. G. M. 2008 Large-eddy simulation of particle-laden channel flow. In *Quality and Reliability of Large-Eddy Simulations* (ed. J. Meyers, B. J. Geurts & P. Sagaut), pp. 367–378. New York: Springer.

KUERTEN, J. G. M., GEURTS, B. J., VREMAN, A. W. & GERMANO, M. 1999 Dynamic inverse modeling and its testing in large-eddy simulations of the mixing layer. *Phys. Fluids* **11** (12), 3778–3785.

KUERTEN, J. G. M. & VREMAN, A. W. 2005 Can turbophoresis be predicted by large-eddy simulation? *Phys. Fluids* **17**, 011701.

KUNDU, P. & COHEN, I. 2004 *Fluid Mechanics*, 3rd edn. San Diego: Academic Press.

LANDAU, L. D. & LIFSHITZ, E. M. 1987 *Fluid Mechanics, Second Edition: Volume 6 (Course of Theoretical Physics)*, 2nd edn. Oxford: Butterworth-Heinemann.

LEONARDO DA VINCI , circa 1500 The notebooks of Leonardo da Vinci, translated by Jean Paul Richter. Seattle.

LIEN, R. C. & D'ASARO, E. A. 2002 The Kolmogorov constant for the Lagrangian velocity spectrum and structure function. *Phys. Fluids* **14** (12), 4456–4459.

LIEN, R. C., D'ASARO, E. A. & DAIRIKI, G. T. 1998 Lagrangian frequency spectra of vertical velocity and vorticity in high-Reynolds-number oceanic turbulence. *J. Fluid Mech.* **362**, 177–198.

LILLY, D. K. 1967 The representation of small-scale turbulence in numerical simulation experiments. In *IBM Scientific Computing Symp. on Environmental Sciences* (ed. H. H. Goldstine), pp. 195–210. Yorktown Heights, NY.

LINK, O., GOBERT, C., MANHART, M. & ZANKE, U. 2008 Effect of the horseshoe vortex system on the geometry of a developing scour hole at a cylinder. In *Proceedings of the 4th International Conference on Scour and Erosion*, pp. 162–168. Tokyo, Japan.

MALIK, N. A. & VASSILICOS, J. C. 1999 A Lagrangian model for turbulent dispersion with turbulent-like flow structure: Comparison with direct numerical simulation for two-particle statistics. *Phys. Fluids* **11** (6), 1572–1580.

MANHART, M. 1995 Umströmung einer Halbkugel in turbulenter Grenzschicht. PhD thesis, Universität der Bundeswehr München.

MANHART, M. 2004 A zonal grid algorithm for DNS of turbulent boundary layers. *Comput. Fluids* **33** (3), 435–461.

MANHART, M. & FRIEDRICH, R. 2002 DNS of a turbulent boundary layer with separation. *Int. J. Heat Fluid Flow* **23** (5), 572–581.

MANHART, M., TREMBLAY, F. & FRIEDRICH, R. 2001 MGLET: a parallel code for efficient DNS and LES of complex geometries. In *Parallel Computational Fluid Dynamics 2000* (ed. C. B. Jenssen, T. Kvamdal, H. I. Andersson, B. Pettersen, A. Ecer, J. Periaux, N. Satofuka & P. Fox). Amsterdam: Elsevier Science B.V.

MATHIEU, J. & SCOTT, J. 2000 *An Introduction to Turbulent Flow*. Cambridge: Cambridge University Press.

MAXEY, M. R. 1987 The gravitational settling of aerosol particles in homogeneous turbulence and random flow fields. *J. Fluid Mech.* **174**, 441–465.

MAXEY, M. R. & RILEY, J. J. 1983 Equation of motion for a small rigid sphere in a nonuniform flow. *Phys. Fluids* **26** (4), 883–889.

MENEVEAU, C. & LUND, T. S. 1997 The dynamic Smagorinsky model and scale-dependent coefficients in the viscous range of turbulence. *Phys. Fluids* **9** (12), 3932–3934.

MENEVEAU, C., LUND, T. S. & CABOT, W. H. 1996 A Lagrangian dynamic subgrid-scale model of turbulence. *J. Fluid Mech.* **319**, 353–385.

MEYER, D. W. & JENNY, P. 2004 Conservative velocity interpolation for PDF methods. *Proc. Appl. Math. Mech.* **4** (1), 466–467.

MINIER, J. P., PEIRANO, E. & CHIBBARO, S. 2004 PDF model based on Langevin equation for polydispersed two-phase flows applied to a bluff-body gas-solid flow. *Phys. Fluids* **16** (7), 2419–2431.

MITTAL, K. L., ed. 1993 *Particles in Gases and Liquids 3: Detection, Characterization, and Control*, 1st edn. Springer.

MONAGHAN, J. 1985 Extrapolating B splines for interpolation. *J. Comput. Phys.* **60** (2), 253–262.

MORDANT, N., METZ, P., MICHEL, O. & PINTON, J. F. 2001 Measurement of Lagrangian velocity in fully developed turbulence. *Phys. Rev. Lett.* **87** (21), 214501.

NGUYEN, A. V. & SCHULZE, H. J. 2003 *Colloidal Science of Flotation*, 1st edn. Boca Raton: CRC Press.

OESTERLÉ, B. & ZAICHIK, L. I. 2004 On Lagrangian time scales and particle dispersion modeling in equilibrium turbulent shear flows. *Phys. Fluids* **16** (9), 3374–3384.

ONSAGER, L. 1945 The distribution of energy in turbulence. *Phys. Rev.* **68** (11-12), 286.

OSEEN, C. W. 1927 *Neuere Methoden und Ergebnisse in der Hydrodynamik*. Leipzig: Akad. Verl.-Ges.

OUELLETTE, N. T., XU, H., BOURGOIN, M. & BODENSCHATZ, E. 2006 Small-scale anisotropy in Lagrangian turbulence. *New J. Phys.* **8** (6), 102.

OVERHOLT, M. R. & POPE, S. B. 1998 A deterministic forcing scheme for direct numerical simulations of turbulence. *Comput. Fluids* **27** (1), 11–28.

PELLER, N., DUC, A. LE, TREMBLAY, F. & MANHART, M. 2006 High-order stable interpolations for immersed boundary methods. *Int. J. Numer. Methods Fluids* .

POPE, S. B. 1983 A Lagrangian two-time probability density function equation for inhomogeneous turbulent flows. *Phys. Fluids* **26** (12), 3448–3450.

POPE, S. B. 2000 *Turbulent Flows*. Cambridge, UK: Cambridge University Press.

PROTHERO, A. & ROBINSON, A. 1974 On the stability and accuracy of one-step methods for solving stiff systems of ordinary differential equations. *Math. Comput.* **28** (125), 145–162.

READE, W. C. & COLLINS, L. R. 2000 Effect of preferential concentration on turbulent collision rates. *Phys. Fluids* **12** (10), 2530–2540.

REEKS, M. W. 2005 On probability density function equations for particle dispersion in a uniform shear flow. *J. Fluid Mech.* **522**, 263–302.

REYNOLDS, A. M. 2003 Superstatistical mechanics of tracer-particle motions in turbulence. *Phys. Rev. Lett.* **91** (8), 084503.

REYNOLDS, O. 1883 An experimental investigation of the circumstances which determine whether the motion of water shall be direct or sinuous, and of the law of resistance in parallel channels. *Philos. Trans. R. Soc. London* **174**, 935–982.

REYNOLDS, O. 1895 On the dynamical theory of incompressible viscous fluids and the determination of the criterion. *Philos. Trans. R. Soc. London, Ser. A* **186**, 123–164.

REYNOLDS, W. 1990 The potential and limitations of direct and large eddy simulations. In *Whither Turbulence? Turbulence at the Crossroads* (ed. J. L. Lumley), pp. 313–343.

RICHARDSON, L. F. 1922 *Weather Prediction by Numerical Process*. Cambridge: Cambridge University Press.

RILEY, J. J. 1971 . PhD thesis, Baltimore, Maryland.

RODEAN, H. C. 1991 The universal constant for the Lagrangian structure function. *Phys. Fluids A* **3** (6), 1479–1480.

ROSENBROCK, H. H. 1963 Some general implicit processes for the numerical solution of differential equations. *Comput. J.* **5** (4), 329–330.

ROTTA, J. C. 1972 *Turbulente Strömungen. Eine Einführung in die Theorie und ihre Anwendung*. Stuttgart: Teubner Verlag.

ROUSON, D. W. I. & EATON, J. K. 2001 On the preferential concentration of solid particles in turbulent channel flow. *J. Fluid Mech.* **428**, 149–169.

RUELLE, D. 1986 Resonances of chaotic dynamical systems. *Phys. Rev. Lett.* **56** (5), 405–407.

RUELLE, D. 2003 *Chaotic evolution and strange attractors*. Cambridge: Cambridge University Press.

SAGAUT, P. 2006 *Large Eddy Simulation for Incompressible Flows*. New York: Springer.

SAWFORD, B. L. 1991 Reynolds number effects in Lagrangian stochastic models of turbulent dispersion. *Phys. Fluids A* **3** (6), 1577–1586.

SAWFORD, B. L. 2001 Turbulent relative dispersion. *Ann. Rev. Fluid Mech.* **33**, 289–317.

SCHILLER, L. & NAUMANN, A. Z. 1933 Über die grundlegenden Berechnungen bei der Schwerkraftaufbereitung. *Ver. Deut. Ing.* **77**, 318–320.

SCHLATTER, P. 2004 LES of transitional flows using the approximate deconvolution model. *Int. J. Heat Fluid Flow* **25** (3), 549–558.

SEELEY, L. E., HUMMEL, R. L. & SMITH, J. W. 1975 Experimental velocity profiles in laminar flow around spheres at intermediate Reynolds numbers. *J. Fluid Mech.* **68** (03), 591–608.

SEHMEL, G. A. 1980 Particle and gas dry deposition: A review. *Atmos. Environ.* **14** (9), 983–1011.

SHOTORBAN, B. & BALACHANDAR, S. 2007 A Eulerian model for large-eddy simulation of concentration of particles with small Stokes numbers. *Phys. Fluids* **19** (11), 118107.

SHOTORBAN, B. & MASHAYEK, F. 2005 Modeling subgrid-scale effects on particles by approximate deconvolution. *Phys. Fluids* **17** (8), 081701.

SHOTORBAN, B. & MASHAYEK, F. 2006 A stochastic model for particle motion in large-eddy simulation. *J. Turbul.* **7**, N18.

SHOTORBAN, B., ZHANG, K. & MASHAYEK, F. 2007 Improvement of particle concentration prediction in large-eddy simulation by defiltering. *Int. J. Heat Mass Transfer* **50** (19-20), 3728–3739.

SIMEON, B. 1998 Order reduction of stiff solvers at elastic multibody systems. *Appl. Numer. Math.* **28** (2-4), 459–475.

SIMEON, B. 2001 Numerical analysis of flexible multibody systems. *Multibody Sys. Dyn.* **6** (4), 305–325.

SIMONIN, O., DEUTSCH, E. & MINIER, J. P. 1993 Eulerian prediction of the fluid/particle correlated motion in turbulent two-phase flows. *Appl. Sci. Res.* **51**, 275–283.

SIMONIN, O., ZAICHIK, L. I., ALIPCHENKOV, V. M. & FÉVRIER, P. 2006 Connection between two statistical approaches for the modelling of particle velocity and concentration distributions in turbulent flow: The mesoscopic Eulerian formalism and the two-point probability density function method. *Phys. Fluids* **18** (12), 125107.

SMAGORINSKY, J. 1963 General circulation experiments with the primitive equations. *Mon. Weather Rev.* **91** (3), 99–164.

SOO, S. L. 1975 Equation of motion of a solid particle suspended in a fluid. *Phys. Fluids* **18** (2), 263–264.

SQUIRES, KYLE D. & EATON, JOHN K. 1990 Particle response and turbulence modification in isotropic turbulence. *Physics of Fluids A: Fluid Dynamics* **2** (7), 1191–1203.

SQUIRES, K. D. & EATON, J. K. 1991 Preferential concentration of particles by turbulence. *Phys. Fluids* **3**, 1169–1178.

SREENIVASAN, K. R. 1995 On the universality of the Kolmogorov constant. *Phys. Fluids* **7** (11), 2778–2784.

STOER, J. & BULIRSCH, R. 2002 *Introduction to Numerical Analysis*, 3rd edn. New York: Springer.

STOLZ, S. & ADAMS, N. A. 1999 An approximate deconvolution procedure for large-eddy simulation. *Phys. Fluids* **11** (7), 1699–1701.

STOLZ, S., ADAMS, N. A. & KLEISER, L. 2001a An approximate deconvolution model for large-eddy simulation with application to incompressible wall-bounded flows. *Phys. Fluids* **13** (4), 997–1015.

STOLZ, S., ADAMS, N. A. & KLEISER, L. 2001b The approximate deconvolution model for large-eddy simulations of compressible flows and its application to shock-turbulent-boundary-layer interaction. *Phys. Fluids* **13** (10), 2985–3001.

STONE, H. L. 1968 Iterative solution of implicit approximations of multidimensional partial differential equations. *SIAM J. Num. Anal.* **5** (3), 530–558.

STRICHARTZ, R. S. 1994 *A Guide to Distribution Theory and Fourier Transforms*, 1st edn. Boca Raton: CRC-Press.

SULLIVAN, N. P., MAHALINGAM, S. & KERR, R. M. 1994 Deterministic forcing of homogeneous, isotropic turbulence. *Phys. Fluids* **6** (4), 1612–1614.

SUNDARAM, S. & COLLINS, L. R. 1999 A numerical study of the modulation of isotropic turbulence by suspended particles. *J. Fluid Mech.* **379**, 105–143.

TANEDA, S. 1956 Experimental investigation of the wake behind a sphere at low Reynolds numbers. *J. Phys. Soc. Jpn.* **11**.

TANG, L., WEN, F., YANG, Y., CROWE, C. T., CHUNG, J. N. & TROUTT, T. R. 1992 Self-organizing particle dispersion mechanism in a plane wake. *Phys. Fluids* **4**, 2244–2251.

TAYLOR, G. I. 1922 Diffusion by continuous movements. *Proc. Lon. Math. Soc.* **s2-20**, 196–212.

TAYLOR, G. I. 1953 Dispersion of soluble matter in solvent flowing slowly through a tube. *Proc. R. Soc. London, Ser. A* **219** (1137), 186–203.

TAYLOR, G. I. 1954 The dispersion of matter in turbulent flow through a pipe. *Proc. R. Soc. London, Ser. A* **223** (1155), 446–468.

TAYLOR, M. A., KURIEN, S. & EYINK, G. L. 2003 Recovering isotropic statistics in turbulence simulations: The Kolmogorov 4/5th law. *Phys. Rev. E* **68** (2), 026310.

TCHEN, C. M. 1947 Mean value and correlation problems connected with the motion of small particles suspended in a turbulent fluid. PhD thesis, University of Delft.

TENNEKES, H. & LUMLEY, J. L. 1972 *A First Course in Turbulence*. Cambridge, MA: MIT Press.

TOSCHI, F. & BODENSCHATZ, E. 2009 Lagrangian properties of particles in turbulence. *Ann. Rev. Fluid Mech.* **41** (1), 375–404.

TOWNSEND, A. A. R. 1975 *The Structure of Turbulent Shear Flow*, 2nd edn. Cambridge: Cambridge University Press.

UHLMANN, M. 2008 Interface-resolved direct numerical simulation of vertical particulate channel flow in the turbulent regime. *Phys. Fluids* **20** (5), 053305+.

VAILLANCOURT, P. A., YAU, M. K., BARTELLO, P. & GRABOWSKI, W. W. 2002 Microscopic approach to cloud droplet growth by condensation. Part II: Turbulence, clustering, and condensational growth. *J. Atmos. Sci.* **59** (24), 3421–3435.

VREMAN, A. W. 2007 Turbulence characteristics of particle-laden pipe flow. *J. Fluid Mech.* **584**, 235–279.

VREMAN, B., GEURTS, B., DEEN, N., KUIPERS, J. & KUERTEN, J. 2009 Two- and four-way coupled Euler-Lagrangian large-eddy simulation of turbulent particle-laden channel flow. *Flow Turbul. Combust.* **82** (1), 47–71.

VREMAN, B., GEURTS, B. & KUERTEN, J. G. M. 1997 Large-eddy simulation of the turbulent mixing layer. *J. Fluid Mech.* **339**, 357–390.

WALPOT, R. J. E., VAN DER GELD, C. W. M. & KUERTEN, J. G. M. 2007 Determination of the coefficients of Langevin models for inhomogeneous turbulent flows by three-dimensional particle tracking velocimetry and direct numerical simulation. *Phys. Fluids* **19** (4), 045102.

WANG, L. P. & MAXEY, M. R. 1993 Settling velocity and concentration distribution of heavy particles in homogeneous isotropic turbulence. *J. Fluid Mech.* **256**, 27–68.

WANG, L. P. & STOCK, D. E. 1993 Dispersion of heavy particles by turbulent motion. *J. Atmos. Sci.* **50**, 1897–1913.

WANG, L. P., WEXLER, A. S. & ZHOU, Y. 2000 Statistical mechanical description and modelling of turbulent collision of inertial particles. *J. Fluid Mech.* **415**, 117–153.

WANG, Q. & SQUIRES, K. D. 1996 Large eddy simulation of particle-laden turbulent channel flow. *Phys. Fluids* **8** (5), 1207–1223.

WILLIAMSON, J. H. 1980 Low-storage Runge-Kutta schemes. *J. Comput. Phys.* **35**, 48–56.

YAMAMOTO, Y., POTTHOFF, M., TANAKA, T., KAJISHIMA, T. & TSUJI, Y. 2001 Large-eddy simulation of turbulent gas-particle flow in a vertical channel: effect of considering inter-particle collisions. *J. Fluid Mech.* **442**, 303–334.

YANG, C. Y. & LEI, U. 1998 The role of the turbulent scales in the settling velocity of heavy particles in homogeneous isotropic turbulence. *J. Fluid Mech.* **371**, 179–205.

YANG, Y., HE, G. W. & WANG, L. P. 2008 Effects of subgrid-scale modeling on Lagrangian statistics in large-eddy simulation. *J. Turbul.* **9**, N8.

YEUNG, P. K. & POPE, S. B. 1988 An algorithm for tracking fluid particles in numerical simulations of homogeneous turbulence. *J. Comput. Phys.* **79** (2), 373–416.

YOSHIZAWA, A. 1982 A statistically-derived subgrid model for the large-eddy simulation of turbulence. *Phys. Fluids* **25** (9), 1532–1538.

Appendix: Computational requirements for SOI

Chapter 7 showed that SOI performs well in terms of accuracy. In the present chapter the numerical costs for SOI are computed. It is shown that the number of floating point operations with SOI is only 7% higher than for fourth order interpolation without particle-LES model.

The preprocessing step for SOI consists mainly of 1D and 2D optimization. Thus, the computational costs in preprocessing are negligible in comparison to the 3D simulation.

At runtime, SOI consists of interpolation only. It should be remarked that the number of nodes for the splines $\hat{w}_l$ and $\hat{w}_t$ merely affect the computational costs of the method. If the number of nodes is increased then only the memory requirements for storing the spline coefficients will increase slightly. However, regarding the overall memory requirements, the memory for storing the splines is negligible.

In the following, a four point stencil for SOI and a five point stencil for ADM will be assumed. This corresponds to the implementations of the previous chapters.

Table A.1 gives an overview on the computational costs with ADM, stochastic models and SOI. Where interpolation is necessary, always fourth order interpolation is assumed. N_p denotes the number of particles, N^3 the number of grid points. In the following, these costs are computed in detail.

Table A.1: Computational costs for computation of fluid velocity at particle position in terms of floating point operations (flops). N_p denotes the number of particles, N^3 the number of grid points. Where interpolation is necessary, always fourth order interpolation is assumed.

	flops
no particle-LES model	$609 N_p$
ADM	$249 N^3 + 609 N_p$
stochastic models	$> 1218 N_p$
SOI	$651 N_p$

Denote in the following the position of a particle by (x, y, z) and enumerate the grid points x_i, y_j, z_k such that

$$x_1 < x_2 < x < x_3 < x_4, \quad y_1 < y_2 < y < y_3 < y_4 \quad \text{and} \quad z_1 < z_2 < z < z_3 < z_4. \quad \text{(A.1)}$$

Appendix: Computational requirements for SOI   169

Computational costs of Lagrangian fourth order interpolation. For Lagrangian fourth order interpolation, one computes first for each particle the distance to the neighbouring grid points,

$$\xi_i = x - x_i, \quad \eta_j = y - y_j, \quad \zeta_k = z - z_k, \quad i,j,k \in \{1,2,3,4\}, \qquad 12 \text{ flops.} \quad (A.2)$$

'flops' stands for floating point operations, a measure for numerical costs. Then, one evaluates the polynomials w_{cub}, cf. section 7.1.1, with a Horner scheme (cf. e.g. Deuflhard & Hohmann, 2008; Stoer & Bulirsch, 2002; Freund & Hoppe, 2007)

$$w_{cub}(\xi_i), \quad w_{cub}(\eta_j), \quad w_{cub}(\zeta_k), \quad i,j,k \in \{1,2,3,4\}, \qquad 12 \cdot 6 = 72 \text{ flops.} \quad (A.3)$$

In the next step one computes $u_{f@p} = \sum_{i,j,k} w_{cub}(\xi_i) w_{cub}(\eta_j) w_{cub}(\zeta_k) u_{i,j,k}$ where $u_{i,j,k}$ denotes the x−component of the fluid velocity (given data) in (x_i, y_j, z_k),

$$u_{f@p} = \sum_{i=1}^{4} w_{cub}(\xi_i) \sum_{j=1}^{4} w_{cub}(\eta_j) \sum_{k=1}^{4} w_{cub}(\zeta_k) u_{i,j,k}, \qquad 147 \text{ flops.} \quad (A.4)$$

In total, this amounts to 231 flops.

For computing the other two velocity components, the weights can be partially recycled but not fully due to the staggered grid. In detail, for the y−component the weights $w_{cub}(\xi_i)$ and $w_{cub}(\eta_j)$ need to be recomputed whereas $w_{cub}(\zeta_k)$ can be retained. For the z−component, only $w_{cub}(\zeta_k)$ must be recomputed. Altogether this results in 609 flops per particle.

Computational costs of SOI. For SOI, one would likewise compute first for each particle the distance to the neighbouring grid points,

$$\xi_i = x - x_i, \quad \eta_j = y - y_j, \quad \zeta_k = z - z_k, \quad i,j,k \in \{1,2,3,4\}, \qquad 12 \text{ flops,} \quad (A.5)$$

then evaluate the cubic splines $\hat{w}_l$ or $\hat{w}_t$ with a Horner scheme (denoted for the first velocity component u_1),

$$\hat{w}_l(\xi_i), \quad \hat{w}_t(\eta_j), \quad \hat{w}_t(\zeta_k), \quad i,j,k \in \{1,2,3,4\}, \qquad 12 \cdot 6 = 72 \text{ flops,} \quad (A.6)$$

divide by the respective sums,

$$\alpha = \sum_{i=1}^{4} \hat{w}_l(\xi_i), \quad \beta = \sum_{j=1}^{4} \hat{w}_t(\eta_j), \quad \gamma = \sum_{k=1}^{4} \hat{w}_l(\zeta_k), \qquad 3 \cdot 3 = 9 \text{ flops} \quad (A.7a)$$

$$w_l(\xi_i) = \frac{\hat{w}_l(\xi_i)}{\alpha}, \quad w_t(\eta_j) = \frac{\hat{w}_t(\eta_j)}{\beta}, \quad w_t(\zeta_k) = \frac{\hat{w}_t(\zeta_k)}{\gamma}, \qquad 3 \cdot 4 = 12 \text{ flops} \quad (A.7b)$$

and compute $u_{f@p} = \sum_{i,j,k} w_l(\xi_i) w_t(\eta_j) w_t(\zeta_k) u_{i,j,k}$,

$$u_{f@p} = \sum_{i=1}^{4} w_l(\xi_i) \sum_{j=1}^{4} w_t(\eta_j) \sum_{k=1}^{4} w_t(\zeta_k) u_{i,j,k}, \qquad 147 \text{ flops.} \quad (A.8)$$

In total, this amounts to 252 flops. For computing the other two velocity components, the weights can again be partially recycled but less than for fourth order interpolation because the stencils in longitudinal and transversal direction differ. Thus, for three dimensions the costs read 651 flops, only 1.07 times the costs for fourth order interpolation.

Thus, the numerical costs of SOI are almost equal to the numerical costs of fourth order interpolation without particle-LES model. From this point of view, modelling with SOI is almost 'for free'.

Computational costs of the stochastic models. For comparison, the computational overhead for the stochastic models stems mainly from interpolation of further terms such as the material derivative, cf. equations (6.5) and (6.8). If these are also interpolated with a fourth order scheme, then the computational costs of the stochastic models are approximately twice as high as the costs for SOI.

Computational costs of ADM. For ADM, the computational costs depend on the number of LES grid points. This means that if the ratio between the number of particles and the number of grid cells is high, then ADM is computationally less expensive than SOI and vice versa. The exact limit depends on the size of the ADM stencil.

For a comparison, assume that the particles are distributed homogeneously over the whole domain. Then, one can compute the limit of the number of particles per cell above which ADM is computationally less expensive than SOI.

Concerning the parameters of both models, assume the same choice with which all computations in this thesis were conducted. In particular, assume a five point ADM stencil, the fourth order Lagrangian interpolation with ADM and the interval $[-2, 2]$ as support of the SOI stencil.

Denote by N_p the number of particles in the domain and by N^3 the number of grid cells. For SOI, the computational costs for interpolation of fluid velocity at particle position read $651 N_p$ flops. For ADM, one needs to multiply for each grid point the neighbouring 5^3 grid points with the ADM coefficients, and add them up. This makes $N^3 \cdot (125 + 124)$ flops. In addition, the interpolation requires $609 N_p$ flops. Concluding, ADM requires less CPU time than SOI if $249 N^3 + 609 N_p < 651 N_p$. This is equivalent to $N_p > 6 N^3$, i.e. if the number of particles per cell is higher than 6. However, this will not occur in a typical application because regarding computational requirements, it is reasonable to limit the number of particles per CPU to 10^6 and the number of cells per CPU to $4 \cdot 10^6$. Thus, if the number of particles per CPU is higher than 6 times the number if grid points, then the CPU time for the particles would outreach the CPU time for the carrier flow by far. All these considerations are only relevant in LES. In LES, one would make the grid as fine as possible, i.e. the CPU time for the particles cannot be significantly higher than the CPU time for the carrier flow. Therefore, in a typical application $N_p < 6 N^3$, i.e. the CPU time for ADM is higher than for SOI.

Die VDM Verlagsservicegesellschaft sucht für wissenschaftliche Verlage abgeschlossene und herausragende

Dissertationen, Habilitationen, Diplomarbeiten, Master Theses, Magisterarbeiten usw.

für die kostenlose Publikation als Fachbuch.

Sie verfügen über eine Arbeit, die hohen inhaltlichen und formalen Ansprüchen genügt, und haben Interesse an einer honorarvergüteten Publikation?

Dann senden Sie bitte erste Informationen über sich und Ihre Arbeit per Email an *info@vdm-vsg.de*.

Sie erhalten kurzfristig unser Feedback!

VDM Verlagsservicegesellschaft mbH
Dudweiler Landstr. 99
D - 66123 Saarbrücken

Telefon +49 681 3720 174
Fax +49 681 3720 1749

www.vdm-vsg.de

Die VDM Verlagsservicegesellschaft mbH vertritt

Printed by Books on Demand GmbH, Norderstedt / Germany